# Science: Philosophy, History and Education

**Scope of the Series**

This book series serves as a venue for the exchange of the complementary perspectives of science educators and HPS scholars. History and philosophy of science (HPS) contributes a lot to science education and there is currently an increased interest for exploring this relationship further. Science educators have started delving into the details of HPS scholarship, often in collaboration with HPS scholars. In addition, and perhaps most importantly, HPS scholars have come to realize that they have a lot to contribute to science education, predominantly in two domains: a) understanding concepts and b) understanding the nature of science. In order to teach about central science concepts such as "force", "adaptation", "electron" etc, the contribution of HPS scholars is fundamental in answering questions such as: a) When was the concept created or coined? What was its initial meaning and how different is it today? Accordingly, in order to teach about the nature of science the contribution of HPS scholar is crucial in clarifying the characteristics of scientific knowledge and in presenting exemplar cases from the history of science that provide an authentic image of how science has been done. The series aims to publish authoritative and comprehensive books and to establish that HPS-informed science education should be the norm and not some special case. This series complements the journal Science & Education http://www.springer.com/journal/11191Book Proposals should be sent to the Publishing Editor at Claudia.Acuna@springer.com

More information about this series at http://www.springer.com/series/13387

Sibel Erduran • Ebru Kaya

# Transforming Teacher Education Through the Epistemic Core of Chemistry

## Empirical Evidence and Practical Strategies

Sibel Erduran
Department of Education
University of Oxford
Oxford, UK

Ebru Kaya
Department of Mathematics and Science
Education
Boğaziçi University
Istanbul, Turkey

ISSN 2520-8594          ISSN 2520-8608   (electronic)
Science: Philosophy, History and Education
ISBN 978-3-030-15328-1          ISBN 978-3-030-15326-7   (eBook)
https://doi.org/10.1007/978-3-030-15326-7

This Springer imprint is published by the registered company Springer Nature Switzerland AG.
The registered company address is: Gewerbestrasse 11, 6330 Cham, Switzerland

*To my mother Ayten Erduran, father Erol Erduran and aunt Suna Erduran for instilling in me a sense of strength through reconciliation and humility.*

*To my mother Guner Kaya, my father Emrah Kaya and my sisters Esra and Ecem for their continued support and love and to my lovely niece Defne.*

# Foreword

This book was conceived during a break at the 2015 ESERA conference in Helsinki, when I suggested to Sibel Erduran to write a book on philosophy of science and science education. Sibel was already thinking about such a book, so my suggestion motivated her to start working on it. And I am very glad that you are now holding it in your hands.

A main goal of this book series is to show not only why history and philosophy of science can make important contributions to science education but also how this can be done. The present book by Sibel Erduran and Ebru Kaya is a fine example of how philosophy of science can enrich and enhance science education, focusing on chemistry and chemistry education. The book actually makes many different contributions, but the authors have chosen two of these to highlight: the importance of teacher education and the practical aspect of using philosophy of science in science education. Let us see why these are very important.

On the one hand, we have teachers and teacher education. Anyone who has taught in a school classroom is aware of the multifactorial and complex nature of teaching. There is so much going on at the same time in any classroom that teachers need to have a wide range of skills and tools available in order to teach effectively. I do not want to diminish the importance of the various factors that affect the quality of education, but I strongly believe that teachers are the most critical one. Having worked as a school teacher myself for 12 years, I know that a teacher has the power to inspire and motivate students, as well as to turn them away of the subject taught, no matter what educational curricula and policies suggest. It is the teacher who can translate the educational policy to classroom practice; it is the teacher who can present the topics of the curriculum to students; and, most importantly, it is the teacher who can arouse students' interest, help them express their preconceptions and guide them to understanding. A well-prepared teacher can achieve these more easily and can teach more effectively than an unprepared one. Therefore, focusing on teacher education, as the present book does, is crucial.

On the other hand, we have the philosophy of science. A major criticism of scholarship that aims at bringing philosophy of science in science education is that this is practically impossible as teachers neither are philosophers of science nor are they supposed to educate the new generations of philosophers of science. Science teaching has to be down-to-earth, focused on the actual science, and not on the disagreements among philosophers about what science is and how it is done; or so the argument goes. In a nutshell, perhaps the scholarship in philosophy of science is too philosophical and not empirical enough in order to be useful to science teaching with real teachers and students. The present book aims to address these concerns, by drawing on "empirical evidence" for the use of philosophy of science in science education and by making "practical strategies" towards this end. It shows how philosophy of science has indeed been used and also makes practical suggestions about how it could be used.

By introducing the concept of "epistemic core", the authors show that philosophical considerations are both inherent in the teaching of chemistry and directly relevant to it. I would argue that this is the case for all science topics, but we may need to wait for other similar books to provide the empirical evidence and the practical strategies for this. On various occasions, I have been paraphrasing Theodosius Dobzhansky who noted that nothing in biology makes sense except in the light of evolution. I might not go as far as to state that nothing in science education makes sense except in the light of philosophy of science. However, I would wholeheartedly argue that science education makes more sense in the light of philosophy of science, and the present book would be one among those that I would give to colleagues who would like to see why.

University of Geneva                                                                Kostas Kampourakis
Geneva, Switzerland                                                                        Series Editor

# Preface

At first glance, this book might appear not chemical enough to the chemist, not philosophical enough to the philosopher and too philosophical to the teacher educator. One has to look closer to recognise that it occupies a very unusual space: the intersection of theoretical themes, empirical evidence and practical approaches. Philosophy of chemistry is, by definition, theoretical in nature. Teacher education is both theoretical and empirical. Teacher education is also a pragmatic and practical endeavour charged with the mission of preparing professional teachers. It has to be responsive to educational policy and teaching practice as governed by national politics and institutional infrastructures. If it is going to be impactful at all, the import of ideas from philosophy of chemistry into teacher education for the development of the chemistry teachers has to consider the actual everyday context of teachers and teacher educators. Such is the tall order of the book: the interplay of theoretical frameworks from vastly different disciplines as well as their application in teacher education practice, engaging teacher educators and pre-service teachers.

Who are the key audiences of this book? Part of the content directly relates to the concerns of philosophers of chemistry, chemists, teacher educators, pre-service teachers as well as researchers in teacher education. Chemists, in some countries, are involved in teacher education, although they may not have had formal training about teacher education or teacher education research. Any of these stakeholders will most likely find that their respective interests are constrained in the content of the book. Each stakeholder operates in a different community that does not necessarily overlap with the others'. For instance, how many chemistry teachers read academic journals dedicated to philosophy of chemistry when they are concerned about what resources to use in their next lesson in preparation for the exams? Do philosophers of chemistry visit university teacher education departments to observe teacher training sessions and where, if at all, their subject matter might actually fit? Which doctoral programmes in chemistry education include coursework on philosophy of chemistry to empower future researchers and lecturers in chemistry and teacher education with epistemological insight?

Our main purpose in writing this book was to bridge traditionally disparate areas of work in order to understand how chemistry education can be improved. When we first began thinking about the book and presenting related papers, some of our ideas were met with resistance. For example, at the biennial IHPST conference in Rio de Janeiro, Brazil, in 2015, one of our colleagues remarked that such an approach is "galaxies away", not recognising the potential of transforming theoretical and philosophical ideas into empirical forms for educational researchers and practical purposes of everyday teachers. At the ICCE conference in Sydney, Australia, in 2018, we observed that chemistry lecturers were so focused on chemistry concepts and procedures that epistemic perspectives on chemistry were considered an irrelevant distraction in education. We wondered ourselves how we can bond philosophical reflection with empirical evidence and practical strategies for improving teacher education. Soon, we recognised that because we are bringing together fairly disparate areas of work, by its very definition, the work needs to be an approximation. It is an approximation to a possible way forward in making chemistry teacher education more epistemic than it currently is. We have to indicate that we are teacher educators and researchers of teacher education who are guided by not only research interests but also pragmatic necessities of the contexts where we teach future teachers. We are deeply aware of the opportunities for conceptual clarification through philosophical reflection, as much as we are aware of the constraints to how theoretical arguments might be implemented at a practical level.

Background on philosophy of chemistry provides us with a rich body of work from which we can draw on to inform chemistry education generally. However, this literature does not tell us how to effectively infuse ideas that will be intelligible or relevant for teacher educators and teachers, or indeed how these ideas may fit or not with the institutionalised teacher education policies of different countries. For the latter, we need to appeal to the research and policy evidence in teacher education as an academic research area as well as pragmatic endeavour of professional training. Hence, while we review the literature on the philosophy of chemistry and illustrate a possible destination for the education of chemistry teachers, we recognise that the outcomes for teacher education may take years beyond pre-service teacher education to materialise. What insight we gather from philosophy of chemistry is most certainly not necessarily applicable to the beginning of the journey where pre-service teachers are much more concerned about more basic demands of pedagogy such as behaviour management. Furthermore, as teacher educators, we are ourselves a product of a system that did not prioritise philosophical content in our own educational preparation, and hence, we need to research and learn about how we can infuse such content in our teaching of teachers. Likewise, save a few potential exceptions, our pre-service teachers would have had practically no exposure to epistemological or other meta-perspectives on chemistry to equip them with the higher-order thinking skills about the subject they are expected to teach. Our beginning point in the teacher education journey, thus, needs to take into account not only the background knowledge of the pre-service teachers but also research evidence on how best to teach pre-service teachers. The journey needs to start with an accessible framework which can be used to initiate metacognitive awareness in pre-service teachers. It is such a journey that we embarked on in producing this book which we hope will help other teacher educators to capitalise on epistemological perspectives in their teaching.

We are thankful to many individuals who have supported the production of the book: Kostas Kampourakis, Bernadette Ohmer and Claudia Acuna from Springer for their continued guidance and support and Erin Peter-Burtons, Julie Luft and three anonymous reviewers for their valuable feedback that helped improve the content. We are grateful to the pre-service teachers who have participated in the research project supported by Bogazici University Research Fund (Grant Number: 10621), as well as Selin Akgun and Busra Aksoz for their help in data collection. Finally, we would like to thank our families and friends for their love and support.

Oxford, UK                                                                                Sibel Erduran
Istanbul, Turkey                                                                              Ebru Kaya

# Introduction

This book is a novel attempt to introduce the epistemic core of chemistry education explicitly in a meaningful way into teacher education. The book contains both the theoretical framing of such an approach and concrete examples of implementation and the possibilities that exist when such seemingly abstract topics are tackled with pre-service teachers. This book can be read as a whole or through the study of individual chapters as each chapter contains meaningful input in its own right.

The two authors are well-placed to present such a book. Sibel Erduran has worked at the interface of education, philosophy and chemistry for more than 20 years, starting with a background in chemistry and moving to philosophy and chemistry at the doctoral level. Her teaching activities have involved teacher education for more than 15 years. Sibel Erduran seldom works alone and has worked with collaborators on five continents. This book is an example of such collaboration. Ebru Kaya, the second author of this book, is fast establishing herself as an independent researcher in the area of argumentation and the nature and philosophy of chemistry.

In the words of the authors, the book aims "to synthesise theoretical perspectives from philosophy of chemistry and teacher education, to explore how such perspectives can be infused into the design and implementation of teacher education programmes with impact on pre-service teachers' understanding of epistemic aspects of chemistry" (p. 14).

In my view, these aims are accomplished most successfully as we are treated firstly to three chapters rich in theoretical perspective, followed by an account of an implementation of these ideas into a teacher education context in the next three chapters. The book concludes with self- reflection by the researchers followed by a consideration of the constraints inherent in such a complex activity and concludes with an elucidation of a framework of epistemic identity for chemistry teacher education.

As a chemical education researcher, interested in pedagogical content knowledge (PCK) (Shulman, 1987) and teacher education, I was drawn by the crucial questions raised and addressed in the book such as "What pedagogical content knowledge do teachers need to have in order to support the learning of epistemic themes in chemistry?" and "How can beginning teachers be equipped with understanding of the epistemic aspects of chemistry?" Beginning teachers do not usually leave initial teacher education equipped with much PCK, and immersion in activities such as those portrayed in the book opens the door to reflective teaching and pedagogical reasoning.

The use of the topic of acids and bases as an example resonated with my experience of teaching the topic to teachers at the postgraduate level. The topic offers an excellent example of evolving models of a useful conceptual division of substances, leading finally to an abstract expression of a dichotomy. At least three levels of conceptualisation of acids and bases are discussed at school level leading to the Lewis conception of acids and bases, usually dealt with only at tertiary level. Below I produce an adapted version of Kolb's table (Kolb, 1978) to summarise the difference between these models. In addition to her three models, I have added a fourth, as identified by Halstead (2009).

| Theory | Operational (macro theory) | Arrhenius (water-ion theory) | Brønsted (proton theory) | Lewis (electronic theory) |
|---|---|---|---|---|
| Acid definition | pH <7, taste sour | Provider of $H^+$ in water | Proton donor | Electron pair acceptor |
| Base definition | pH >7, bitter, soapy | Provider of $OH^-$ in water | Proton acceptor | Electron pair donor |
| "Neutralisation" | React chemically to give neutral solution | Formation of water | Proton transfer | Coordinate covalent bond formation |
| Equation | None | $H^+ + OH^- \rightarrow H_2O$ | $HA + B \rightarrow BH + A$ | $A +: B \rightarrow A:B$ |
| Limitations | Everyday only | Water solutions only | Proton transfer reactions only | Generalised theory |

These models of acid-base theory are ideal vehicles for teaching the epistemic core of chemistry using content that is largely taught at secondary school level including the epistemic value of simplicity, referred to as "Ockham's razor", meaning that the simplest applicable model is the most elegant and the best. Using Halstead's ideas, I have added an even simpler theory to Kolb's table, the "operational theory", which operates at the phenomenological level and is used to teach acids and bases in the lower grades of the school. In the book, Erduran and Kaya call on a variety of useful and simple tools which illustrate the use and applicability of visual tools to make accessible in teaching pre-service teachers about the epistemic core of chemistry such as the Benzene Ring Heuristic and the theories, laws and models (TLM) approach to understanding the growth of scientific knowledge. These approaches make a conceptual counter to the traditionally taught "scientific method".

As a PCK researcher, I appreciate the attempt in the book to link knowledge about the nature of science to existing models of teacher knowledge, particularly those around PCK. Researchers in NOS have experimented with combining the constructs of PCK and NOS, but critiques challenge them to identify the "C" or content in NOS. The authors in this book make important links with PCK and NOS, in particular Schwab's (1964) argument for dual knowledge of a domain, the content and the epistemology and the reference by Magnusson, Krajcik and Borko (1999) to the idea of orientations to science teaching. The authors convinced me through the data presented in Chaps. 5 and 6 that they have a highly developed "knowledge for teaching" NOS. Their data also showed that they could draw rich analogies from the pre-service teachers, including those who were not necessarily the best academic performers in the group. The constructs of pedagogical reasoning and reflection are evident in both the design of the intervention and the authors' own self-study of the process.

The book produces useful products. First among these is a possible structure for integrating the teaching of the epistemic core of chemistry in teacher education. Second, they provide a model for the development of pre-service teachers' epistemic identity and a corresponding model for teacher professional development.

Erduran and Kaya acknowledge a number of limitations and challenges in the research described in this book. First among these is the lack of insight into how and if the ideas learnt in the intervention carried over into their teaching. The second is a challenge to others thinking of implementing these ideas. The comparison of the two universities of the authors, Bogazici and Oxford, in two countries show stark comparisons on the availability of opportunity to implement the kind of approach used. The authors certainly show that such approaches are possible and fruitful and despite the challenges can produce substantial outcomes.

University of the Witwatersrand                                    Marissa Rollnick
Johannesburg, South Africa

# References

Halstead, S. E. (2009). *A critical analysis of research done to identify conceptual difficulties in acid-base chemistry.* (MSc), University of KwaZulu-Natal, Pietermaritzburg. Retrieved from file:///C:/Users/09001010/Documents/Research/Acids%20and%20bases%20Halstead_Sheelagh_Edith_2009.pdf

Kolb, D. (1978). Acids and bases. *Journal of Chemical Education, 55*(7), 459–464.

Magnusson, S., Krajcik, J., & Borko, H. (1999). Nature sources and development of pedagogical content knowledge for science teaching. In J. G. Newsome & N. G. Lederman (Eds.), *Examining pedagogical content knowledge: The construct and its implications for science education* (pp. 95–132). Dordrecht, the Netherlands: Kluwer Academic.

Schwab, J. J. (1964). The structure of the disciplines: meanings and significances. In G. W. Ford & L. Pugno (Eds.), *The structure of knowledge and the curriculum* (pp. 6–30). Chicago: Rand McNally.

Shulman, L. S. (1987). Knowledge and teaching: foundations of the new reform. *Harvard Educational Review, 57*(1), 1–22.

# Contents

# List of Figures

# About the Authors

**Sibel Erduran** is a Professor of science education and a Fellow of St Cross College at the University of Oxford, United Kingdom. She also holds Visiting Professorships at University of Oslo, Norway and Zhejiang Normal University, China. She is an Editor for *International Journal of Science Education*, Section Editor for *Science Education,* and serves on the Executive Board of the European Science Education Research Association. Her work experience includes positions in the USA (University of Pittsburgh), Ireland (University of Limerick), as well as other universities in the UK (King's College London and University of Bristol). Previously she held Visiting Professorships at Oxford Brookes University, National Taiwan Normal University, Kristianstad University, Sweden, and Bogazici University, Turkey. Her higher education was completed in the USA at Vanderbilt (Ph.D., science education and philosophy), Cornell (M.Sc., food chemistry and chemistry), and Northwestern (B.A., biochemistry) Universities. She has worked as a chemistry teacher in a high school in northern Cyprus. Her research interests focus on the infusion of epistemic practices of science in science education, in particular chemistry education. Her work has received international recognition through awards from NARST and EASE, and has attracted funding from a range of agencies including the European Union, Gatsby Foundation, and Science Foundation Ireland. She is currently the Principal Investigator of a project on the assessment of practical science funded by the Wellcome Trust, Gatsby Foundation and Royal Society and her edited book on *Argumentation in Chemistry Education* has been published in 2019 by the Royal Society of Chemistry.

**Ebru Kaya** is an Associate Professor of science education at Bogazici University, Turkey. She received her Ph.D. from Middle East Technical University, Turkey, in 2011. Dr. Kaya has been a Visiting Scholar at the University of Bristol, UK, in 2009 and at National Taiwan Normal University, Taiwan, in 2016 and 2017. Her research interests include argumentation and nature of science in science education. Dr. Kaya is the recipient of the Young Scientist Award from Science Academy in Turkey in 2015; the Outstanding Paper Award from East-Asian Association for Science Education (EASE) in Japan in 2016; and the Young Scientist Award from TUBITAK

(Turkish Scientific and Technological Research Council) in 2017. Dr. Kaya has participated in research projects funded by TUBITAK (Turkish Scientific and Technological Research Council) and NARST, and conducted professional development workshops for science teachers in Turkey, Rwanda, Lebanon, and Taiwan. She served as a board member in Turkish Science Education and Research Association from 2012 to 2014, and a committee member in NARST Outstanding Paper Award Committee from 2013 to 2015. Currently Dr. Kaya is a Reviewer for numerous journals including *International Journal of Science Education* and *Science & Education*. Dr. Kaya has been the Principal Investigator of the projects entitled "Nature of Science in Science Teacher Education: A Comparative Research and Development Project" and "University Students' Understanding of Reconceptualized Family Resemblance Approach to Nature of Science: A Case Study" funded by Bogazici University Research Fund.

# Chapter 1
# Philosophy of Chemistry and Chemistry Education

## 1.1  Introduction

In a review of chemistry education research trends from 2004 to 2013, Teo, Goh, and Yeo (2014) reported findings from a content analysis of chemistry education research papers published in two top-tiered chemistry education journals, *Chemistry Education Research and Practice* and *Journal of Chemical Education*, and four top-tiered science education journals, *International Journal of Science Education*, *Journal of Research in Science Teaching*, *Research in Science Teaching* and *Science Education*. The authors observed that of 650 papers in these journals, only (a) 4 or 0.6% were dedicated to history, philosophy and nature of chemistry, and (b) 54 or 8.3% involved pre-service teachers as participants. These results indicate that there is insufficient attention to empirical studies on the infusion of nature of chemistry in chemistry education in general and in the context of pre-service teacher education in particular. Although Teo and colleagues' study did not consider journals that particularly target history and philosophy of chemistry (HPC) (e.g. *Science & Education*, *Foundations of Chemistry*, *HYLE*), the extent of empirical studies in chemistry education research related to the HPC themes is likely to be relatively low in those journals given their overall theoretical scope. Furthermore, although there is a handful of books in the field which the journal content analysis would not have captured (e.g. Niaz, 2011), the research trends suggest that the intersection of the fields of philosophy of chemistry and pre-service teacher education is virtually inexistent.

The lack of empirical studies in the chemistry education research literature focusing on HPC reflects a fairly low inclusion of history and philosophy of science (HPS) content in teacher preparation programmes in the first place. As a key contributor to debates on the inclusion of HPS in science education, Matthews (2014) laments about the poor health of science teacher education:

© Springer Nature Switzerland AG 2019
S. Erduran, E. Kaya, *Transforming Teacher Education Through the Epistemic Core of Chemistry*, Science: Philosophy, History and Education,
https://doi.org/10.1007/978-3-030-15326-7_1

*Teacher education, and specifically the discipline of science education, is not in good philo-
sophical health. Despite all of the concerns and arguments that have long been known and
that have been documented in this book, competence in philosophy and more specifically
HPS is rare in Schools of Education, nor is their attainment much encouraged. In 1989 only
four of fifty-five institutions providing science teacher training in Australia offered any
HPS- related course. In 1990, of the fifteen leading centres of science teacher training in the
US, only half required a course in philosophy of science; the proportion in the remaining
hundreds of centres was far lower (Loving 1991). The situation in the rest of the world is no
more encouraging. Thus, a teacher's grasp of HPS is largely picked up in their own science
courses; and it is seldom consciously examined or refined. This is epistemology by osmosis,
and is less than desirable for the formation of something so influential in teaching practice,
and so important for professional development.* (p.423)

The case of chemistry teacher education faces additional constraints because of the
limited work on philosophy of chemistry itself until fairly recently. Within philosophy of science, chemistry has traditionally had a peripheral existence (Scerri, 2000;
van Brakel, 2000). As Good (1999) explains, chemistry had minimal existence
within philosophy of science primarily due to lack of interest on the part of chemists
themselves:

*One of the characteristics of chemists is, that most have no interest in the philosophy of
science...The disinterest appears to work in both directions. Modern philosophers very seldom give even a passing mention to modern chemical issues (Michael Polanyi and Rom
Harré are among the few exceptions I know of). Recently, a few philosophers have attempted
to discuss 'scientific practice'; but generally they have not included chemical practice. It is
as if philosophers have believed that the way physics is 'done' was the way that all science
is, or should be, done. (Physicists, no doubt, are the source of this opinion.)* (Good, 1999,
pp. 65–66)

The disinterest in the philosophical aspects of chemistry by philosophers, chemists
and educators mirrors earlier observations about a similar lack of interest regarding
history of chemistry, despite some exceptional and useful contributions in chemistry
education (e.g. Chamizo, 2014; Niaz, 2016):

*Chemists, compared with other scientists, have relatively little interest in the history of their
own subject. This situation is reflected, and perpetuated, by the anti-historical character of
most chemical education.* (Stephen Brush quoted by Kauffman, 1989, p. 81)

With a backdrop of a fairly thin background on HPC in chemistry education, it is
worthwhile to investigate briefly what chemistry educators have been advocating in
terms of what chemistry teaching should include as well as what indicative curriculum development efforts have taken place internationally. Both of these issues
regarding the content of chemistry teaching and curriculum have been widely
reported (e.g. Coenders, Terlouw, Dijkstra, & Pieters, 2010) along with evaluations
of the status chemistry education research more broadly (e.g. Cooper, 2018). Hence,
the intention in the next sections is not to provide an exhaustive overview but rather
to provide a selection of themes to contextualise the primary focus of this book
which is the inclusion of epistemic perspectives in chemistry teacher education.

## 1.2 Arguments About Chemistry Teaching

Over 70 years ago, Standen boldly claimed that there are three ways of teaching chemistry or any other science: forwards, backwards and "heuristically". He considers the backward method to be popular at the time and describes it as follows:

> *We start with the structure of the nucleus, of atoms, molecules, ions, and go on from there. Much of the traditional descriptive material-properties of specific elements, specific compounds, industrial processes-must of necessity be ignored or, at best, covered hastily, to give time for a more comprehensive presentation of the major concepts of chemistry, the nature of chemical bonds, the kinetic basis of chemical reactions (collision theory, activation energy, the effect of size and shape of molecules, the statistical nature of reaction, etc.). Such a method can be described as 'backward' because it reverses the true scientific order, which is from facts to theories, and makes a presentation from theories to facts.* (Standen, 1948, p. 506)

He equates the historical approach to the "heuristic method" and is influenced by the early works of his contemporary Conant (1947) who had subsequently developed the *Harvard Case Histories in Experimental Science* (Conant, 1948). Despite his preoccupation with mechanisms of progression in chemistry that would be appropriate for educational purposes, Standen's account is fairly focused on chemistry concepts such as atoms, molecules and bonds as the drivers of teaching, rather than, say the characteristics of the circumstances that drove knowledge development such as criteria for taxonomic characterisation of substances. He continues to make reference to a set of priorities that may drive choices around topics to be taught in the chemistry classroom:

> *The order of priority that we have now arrived at (which may, for reference purposes, be called "epistemological," although I propose to use this unpleasant ten dollar word as little as possible), is quite distinct, in idea, from the historical order, although in practice it may follow the historical order more or less closely.* (Standen, 1948, p.506)

His notion of "order of priority" relates to decision-making about what to teach, and the relationship between experimental results and theories. The reference to "epistemological" priorities is not unpacked in a precise sense. However, his use of the word "epistemological" in relation to chemistry education is fairly unique even relative to contemporary accounts. Despite his reference to "epistemological" and "historical" accounts, Standen (1948) himself is rather sceptical about the infusion of HPC in chemistry education. While he appreciates history of chemistry for its own sake, he does not see value in importing historical episodes from chemistry for educational use:

> *These difficulties were genuine in their time, but it is not necessary to go through them in teaching. In introductory courses we do not teach the battle now in progress; we teach the ground already won, and there the situation is peaceful. It is not dynamic, it is static. To describe it as static sounds almost like a smear word nowadays, when everything has to be "dynamic," but the description is true. The content of introductory courses is largely in the quiet area of science. The formula $H_2O$ has not changed for 90 years and is not likely to change further.* (p.509)

It is quite remarkable that 70 years since Standen's (1948) views of chemistry teaching, the spirit of resistance to the shaping of chemistry education through HPC is still fairly dominant. Although some progress has been made in maintaining the arguments for the inclusion of HPC in chemistry education (e.g. Kaya & Erduran, 2013; Niaz, 2016), at a practical level in schooling, chemistry education continues to fail its students because it presents a version of chemistry that is devoid of chemistry's historical and philosophical content and context.

What do we mean when we say chemistry education continues to fail its students because it presents a version of chemistry that is devoid of chemistry's epistemic content and context? In order to unpack this sentence, let's think about what understanding chemistry can potentially involve. One way of thinking about understanding chemistry is in terms of chemistry's conceptual and methodological content. Here, as an educator, one would be concerned with how students make sense of particular concepts, such as "molecule", "equilibrium" and "bond", and methods such as "fractional distillation" and "titration". This account is very similar to Standen's (1948) approach. Regardless of the teaching approaches emphasised, his perspective was still fairly focused on concepts or put another way, "subject knowledge" of chemistry. This perspective has pervaded much of what has been carried out as chemical education research for a number of decades, producing a vast amount of research (Gabel & Bunce, 1984) although the emphasis could at times be on particular skills such as problem-solving (e.g. Lythcott, 1990) and misconceptions (e.g. Ross & Munby, 1991).

Another account of chemistry education could position chemistry as a domain that produces knowledge and defines educational goals relative to the knowledge characteristics and knowledge production processes of chemistry. In other words, the nature of chemical knowledge and mechanisms of knowledge development are considered from a "birds-eye view" and imported as learning goals and outcomes. In this scenario, teachers and students ask questions about what concepts and methods of chemistry are; how they are generated; how they are adopted or abandoned in time; and what standards and criteria drive knowledge processes in chemistry. Are bonds the same kind of knowledge as knowledge of reaction mechanisms? How do we know there are bonds anyway? How did we get to know about bonds? This second orientation to understanding chemistry places an emphasis on the epistemic nature of chemistry.

The contrast does not imply that chemistry education should consist of the first or the second orientation. These orientations are not necessarily mutually exclusive. Ultimately, it is ideal for learners of chemistry to understand not only the concepts and processes of chemistry but how these aspects of chemistry arise and how they are justified in the first place. However, for research purposes, chemistry education researchers may wish to emphasise different aspects of chemistry to gain in-depth understanding. In the same way that much research on chemistry education have prioritised certain aspects of chemistry (e.g. concepts) and de-emphasised others (e.g. history of chemistry) so that meaningful results can be obtained for particular research questions, researchers can focus on epistemic aspects while setting aside others. Ultimately, the various approaches can collectively contribute to how best to organise chemistry education practice to optimise teaching and learning.

The emphasis on the epistemic content of science in school science is not new. For the past few decades, a vast amount of research literature on history and philosophy of science in science education has been generated (e.g. Duschl, 1990; Hodson, 1988; Klopfer, 1969; Matthews, 2014) that addresses precisely the epistemic aspects of science in relation to questions about science education. Many chemistry educators have investigated how history and philosophy of chemistry can help inform and improve chemistry education (e.g. Chamizo, 1992; Niaz, 1988). Related areas of research such as science studies (e.g. Duschl, Erduran, Grandy, & Rudolph, 2006), epistemic practices (e.g. Kelly, 2011; Sandoval, 2005), argumentation (e.g. Erduran & Jimenez-Aleixandre, 2007) and modelling (Gilbert & Boulter, 2000) are also prevalent in science education research. Among the school science subjects, however, chemistry continues to be a relatively uninterrogated domain in terms of its epistemic aspects. Furthermore, relatively fewer studies have focused on the educational applications of this kind of research (e.g. Sjöström, 2013; Vesterinen, Aksela & Lavonen, 2013; Vilches & Gil-Perez, 2013). A brief survey of developments in the chemistry curriculum will help illustrate further the fairly minimal attention that epistemic aspects of chemistry have received in curricula.

## 1.3   Chemistry Curriculum Development: A Brief Overview

Conventional approaches in chemistry curricula have emphasised content knowledge (e.g. atomic structure and chemical reactions), practical work (e.g. titration experiments, measurement of reaction rates) and societal aspects of chemistry (e.g. effects of chemical pollution on the environment). Globally, many curriculum reform efforts have been based on these approaches. A brief reflection on the history of the chemistry curriculum will illustrate that many reform efforts have been guided by different emphases on the subject knowledge of chemistry. Interdisciplinary Approach to Chemistry (IAC), a 1970s curriculum in the USA for students of 15 or 16, is an example of the "substances approach". The curriculum introduced substances as either mixtures or pure and distinguished between physical and chemical changes. The "atomic structure" approach was used to develop the materials of CHEM Study, a project of the National Science Foundation for grades 10 or 11 students in the USA. The curriculum was widely used in the 1960s and 1970s in a number of countries, either directly or in adapted forms that often also spread the learning over several of the later secondary years. In the 1960s in England, Nuffield "O" Level Chemistry promoted the "chemical reactions" approach. The curriculum began with the problems of separating pure substances from the mixtures and compounds in which they occur naturally. Physical separation processes were used as techniques to solve problems of chemical reactions. The curriculum then moved straight into reactions in which substances are decomposed by heating in air or burned in air to react with oxygen. Chemistry in the Community (ChemCom) curriculum, developed in the USA during the 1980s and 1990s and supported by the National Science Foundation and the American Chemical Society, is exemplary of a societal approach to curriculum design.

With the backdrop of such curriculum development efforts, contemporary arguments have begun to advocate more nuanced accounts of chemistry in the curriculum. Gilbert (2006) highlighted the importance of context in chemistry education. He argued that students should "...be able to provide meaning to the learning of chemistry; they should experience their learning as relevant to some aspect of their lives and be able to construct coherent 'mental maps' of the subject" (p. 960). Gilbert described four models of context and explained that there may be a steady progression across these models. In Model 1, applications are solely used to illustrate the significance of disciplinary concepts. In Model 2, contexts are not conceived as static constructs to which chemical knowledge is applied, but rather they actively affect the meaning attributed to the concepts. Model 3 is characterised by the active involvement of the learner in giving meaning to the content in relevant contexts. Models 2 and 3 can be placed at the sociochemistry level. Finally, in Model 4, the social dimension of context becomes essential as students actively engage in critical reflection. Pedretti and Nazir (2011) proposed to arrange various Science-Technology-Studies accounts from application/design-oriented to historical and logical reasoning-oriented perspectives, focusing on (a) understanding historical embeddedness of disciplinary chemistry, (b) decision-making about complex issues through consideration of empirical evidence and (c) sociocultural, value-centred and socio-ecojustice-oriented and ideologically informed decision-making and actions in chemistry. Hodson (2011) has argued for four levels of sophistication of issues-based science education: (a) appreciating the societal impact of scientific and technological change, (b) recognising that decisions about science and technological development might be made in pursuit of particular interests, (c) developing one's own views and value positions and (d) preparing for and taking actions on socioscientific and environmental issues.

What these example arguments have highlighted was the need to situate chemistry education in its authentic social context. While these examples have been a welcome development in furthering our thinking about how we conceptualise the chemistry curriculum, they have underplayed the philosophical perspectives in relation to chemistry education. There is a scarcity of perspectives that directly target philosophical issues in chemistry education research. One such perspective has been offered by Talanquer (2013) who described ten complementary facets of chemistry knowledge for teaching. These facets are referred to as essential questions, big ideas, crosscutting concepts, conceptual dimensions, knowledge types, dimensional scales, modes of reasoning, contextual issues, historical views and philosophical considerations. As part of the philosophical considerations, Talanquer cites some of the recent work on explorations of how philosophy of chemistry can be applied in chemistry education and provides the examples of chemical laws, models and language as relevant examples. One example is from our own work. In Kaya and Erduran (2013) we applied philosopher Pierre Laszlo's idea of "concept duality" to the analysis of Turkish chemistry curricula, illustrating the relevance of the idea for the chemistry curriculum. For instance, we identified examples from the 11th grade curriculum (i.e. "reduction/oxidation", "strong acid/strong base" and "exothermic change/endo-thermic change") where the "concept duality" theme could be integrated to encourage philosophical considerations of the related concepts.

Another example argument for the inclusion of some philosophical perspectives in the chemistry curriculum was offered by Sjöström (2013) who proposed an enrichment of Mahaffy's model. Mahaffy (2006) suggested a tetrahedron model that includes Johnstone's (1993) chemical triangle at its base. This bottom triangle, also referred to as the chemistry triplet, includes the formal aspects of chemistry teaching: the macroscopic properties and behaviours of chemical substances; the submicroscopic models used to describe, explain and predict chemical properties and phenomena; and the symbolic representations developed to represent chemical concepts and ideas. In Mahaffy's model, the top of the tetrahedron represents the human element, including both relevant contexts and productive practices. Sjöström (2013) has proposed that Mahaffy's tetrahedron could be enriched by recognising different levels of complexity in the analysis of humanistic aspects in chemistry education. These levels may be represented as different layers of the tetrahedron as one moves from the disciplinary bottom triangle towards the humanistic apex.

The first level in this progression, identified as Applied Chemistry, characterises approaches to chemistry education that introduce the human element by focusing on everyday life issues and different applications of chemistry. At a higher level of complexity, labelled sociochemistry, the teaching of chemistry includes approaches aimed at evaluation of the development and uses of chemistry knowledge, practices and products, as well as understanding of the sociocultural embeddedness of scientific work and ideas. At the top of the enriched tetrahedron, "Critical-Reflexive Chemistry" engages students in a reflective analysis of historical, philosophical, sociological and cultural perspectives, as well as in critical-democratic action for socio-ecojustice. In general, the bottom of the tetrahedron is characterised with disciplinary and formal aspects of chemistry. The first level, Applied Chemistry, is characterised with pragmatic aspects of the discipline and the top of the tetrahedron – the second and third levels – with different reflective aspects.

The recent emphasis on the inclusion of philosophical aspects of chemistry in the chemistry curriculum coincides with the formalisation of "philosophy of chemistry" as a discipline of inquiry within philosophy of science almost 20 years ago. As a potentially fruitful territory for curriculum development as well as other aspects of research and practice of chemistry education, we turn to explore the implications of philosophy of chemistry for chemistry education following a brief review of philosophy of chemistry as a field.

## 1.4    Philosophy of Chemistry: A New Source of Information for Chemistry Education

In 1997, Eric Scerri published a short paper with the thought-provoking title, *Are chemistry and philosophy miscible?* (Scerri, 1997). The use of a chemistry metaphor in communicating the bringing together of the seemingly disparate fields – chemistry and philosophy – was insightful, particularly given the earlier observation about disinterest in philosophical aspects of chemistry. Scerri's paper signalled an upsurge of interest in the study of chemistry from a philosophical perspective. An increasing number of books, journals, conferences and associations focused on the articulation of how chemistry could be understood from a philosophical perspective (Baird, Scerri, & McIntyre, 2006; McIntyre & Scerri, 1997; Scerri, 1997, 2000; van Brakel, 1997, 2000, 2010) have surfaced. Consider, for instance, the now established *International Society for the Philosophy of Chemistry* which has recently held a symposium in Leuven, Belgium. Journals such as *HYLE* and *Foundations of Chemistry* have focused exclusively on the philosophical investigations on chemistry. Books such as Eric Scerri's *The Periodic Table: Its Story and Its Significance* have been published that provide collections that interrogate chemistry from a philosophical perspective (Scerri, 2007). The *Stanford Dictionary of Philosophy* has included an entry on philosophy of chemistry (Weisberg, Needham, & Hendry, 2011).

Unfortunately, the same dynamism of scholarship cannot be attributed to the infusion of philosophy of chemistry in chemical education research and practice. The development of new perspectives on how philosophical aspects of chemistry can inform education has had rather slow progress. In *Chemical Education: Towards Research-Based Practice*, Gilbert and colleagues (2003) noted that research on chemical education drawing perspectives from philosophy of chemistry represented "research aimed at generating new knowledge, the impact of which on practice is uncertain, diffuse or long-term" (p. 398). *Science & Education* was one of the first journals to dedicate space to the work of educators preoccupied with the synthesis of perspectives from philosophy of chemistry for application in chemical education (e.g. Erduran 2001, 2005, 2007). A recent edition with contributions from philosophers, chemists and educators (Erduran, 2013) is testament to the journal's vision in pushing boundaries for innovative scholarship, and it illustrates the small but growing interest in capitalising on the philosophical aspects of chemistry for the improvement of chemical education.

The volume edited by Erduran (2013) consists of papers that deal with a range of issues raised in philosophy of chemistry in application to chemical education. One set of papers focused on the nature of chemical knowledge, particularly in relation to models, explanations and laws. Woody (2013) used the ideal gas law as an example in reviewing contemporary research in philosophy of science concerning scientific explanation. She clarifies the inferential, causal, unification and erotetic conceptions of explanation. Chemical laws were the primary focus of Emma Tobin's work (Tobin, 2013). She provided an overview of the laws in chemistry and reflects on the recent debates on the particular and universal nature of laws, concluding that while generalisations in chemistry are diverse and heterogeneous, a distinction between idealisations and approximations can nevertheless be used to successfully taxonomise them. Aduriz-Bravo (2013) challenged the received, syntactic conception of scientific theories and argues for a model-based account of the nature of science. The significance of models and modelling in chemistry was highlighted by Chamizo (2013) who presented a typology of models and their relation to modelling. Izquierdo-Aymerich (2013) argued for the generation of chemical criteria from the history and philosophy of chemistry for informing the design of chemistry curriculum.

Another set of papers focused on particular epistemological themes that have generated a great deal of debate in philosophy of chemistry in recent years. The authors extend these debates to the curricular, textbook and teaching contexts, and in so doing they elaborate on their potential instantiation in education. Newman (2013) covered emergence and supervenience, key concepts related to the micro-macro relationships in chemistry. He provides a model for teaching chemistry with the potential to enhance fundamental understanding of chemistry. Laszlo (2013) argued that chemistry ought to be taught in like manner to a language, on the dual evidence of the existence of an iconic chemical language, of formulas and equations and of chemical science being language-like and a combinatorial art. Universality and specificity of chemistry are explored by Mariam Thalos who argued that chemistry possesses a distinctive theoretical lens – a distinctive set of theoretical concerns regarding the dynamics and transformations of a variety of organic and inorganic substances (Thalos, 2013). While she agrees that chemical facts bear a reductive relationship to physical facts, she argues that theoretical lenses of physics and chemistry are distinct. Manuel Fernandez-Gonzalez (2013) discusses the concept of pure substance, an idealised entity whose empirical correlate is the laboratory product. A common structure for knowledge construction is proposed for both physics and chemistry with particular emphasis on the relations between two of the levels: the ideal level and the quasi-ideal level.

Kaya and Erduran (2013) focused on concept duality, chemical language and structural explanations, to illustrate how chemistry textbooks could be improved with insights from such work. They provide some example scenarios of how these ideas could be implemented at the level of the chemistry classroom. Vicente Talanquer (2013) presented a case that dominant universal characterisations of the nature of science fail to capture the essence of the particular disciplines. The central goal of this position paper was to encourage reflection about the extent to which dominant views about quality science education based on universal views of scientific practices may constrain school chemistry. Activities, practices and values of chemistry are interrogated in a third set of papers. Earley (2013) recommended that chemistry educators shift to a different "idea of nature", an alternative "worldview". Garritz (2013) illustrated how teaching history and philosophy of physical sciences can illustrate that controversies and rivalries among scientists play a key role in the progress of science and why scientific development is not only founded on the accumulation of experimental data. The case of quantum mechanics and quantum chemistry is used as an example because it is historically full of controversies.

Pinto-Ribeiro and Costa-Pereira (2013) illustrated how pluralism in philosophical perspectives can result in different cognitive, learning and teaching styles in chemical education. Their paper reported on the authors' experiences in Portugal in drafting structural ideas and planning for the subject "didactic of chemistry" based on the philosophy of chemistry. Veli-Matti Vesterinen, Maija Aksela and Jari Lavonen assessed how the different aspects of nature of science (NOS) were represented in Finnish and Swedish upper secondary school chemistry textbooks (Vesterinen et al., 2013). They presented an empirical study where dimensions of NOS were analysed from five popular chemistry textbook series. Vilches and Gil-Perez (2013) reflected on the UN Decade of Education for Sustainable Development and how chemical education for sustainability remains practically absent nowadays in many high school and university chemistry curricula all over the world. They explored the belief that genuine scientific activity lies beyond the reach of moral judgement logically. They proposed possible contributions of chemistry and chemical education to the construction of a sustainable future. Sjöström (2013) reported on Bildung-oriented chemistry education, based on a reflective and critical discourse of chemistry. This orientation is contrasted with the dominant type of chemistry education, based on the mainstream discourse of chemistry. Bildung-oriented chemistry education includes not only content knowledge in chemistry but also knowledge about chemistry, both about the nature of chemistry and about its role in society.

Pierre Laszlo (1999) provided a case for the dual and circular nature of concepts in chemistry. Laszlo argued that the nature of duality and circulation of concepts pose a departure from the deductive reasoning that dominates physics, for instance. The circularity of core concepts in chemistry can be a barrier for deductive reasoning in chemistry. Laszlo mentioned dualism to be an inherent characteristic in the development of chemistry via alchemy by giving examples from Chinese, Western and Arab alchemical thinking. He defined dual concepts in chemistry as the concepts used as pair opposites (e.g. stability-instability) and complementary of each other (e.g. acid and base). For example, the acid concept is defined by referencing the base concept and vice versa. The definitions of acid-base dual concepts are detailed through the use of models such as the Arrhenius model, Brønsted-Lowry model and Lewis model. "Oxidant and reductant" are another example of dual concepts. The duality between these two concepts is addressed with electron transfer. A further example is about stability of chemical structures and the attendant energy states. "Stable" is defined as that with the lowest energy, and "unstable" is defined as that with the highest energy. In this example as well, the concepts are defined in relation to each other in terms of energy levels.

There might be other core concepts in the circulation of concepts. The concept of activation includes the "stable-unstable" duality, and the concept of catalysis includes the "inert-labile" duality. The other examples of circularity of concepts are "electrophile-nucleophile", "substrate-reagent" and "receptor-agonist". Given the pattern of concept dualities in chemistry, Laszlo pointed out that chemistry is concerned with the union of opposites. Because of circularity of concepts, chemistry can pose a difficulty for students. For instance, the attempt to interlink the core chemical concepts by circulating them might give rise to misunderstandings or partial understandings. Chemistry with its hybrid character is both a science of matter because of operations, and a science of the mind because of symbolic reasoning. Laszlo suggested that awareness of concept duality could potentially help teachers and students. He proposed that teachers should address the circulation of concepts rather than skipping the issue in chemistry instruction.

In summary, there is now a body of research on philosophy of chemistry that can supplement the work that has already been conducted more generally on the applications of HPS in science education (e.g. Matthews, 2014). Research on philosophy of chemistry can inform both the theory and practice of chemistry education. For example, philosophical accounts of chemistry may point to curriculum content that is not currently in focus that may potentially help improve teaching and learning. For example, the explicit illustration of Laszlo's "duality" theme could be included as a learning goal that may consolidate understanding of disparate concepts.

## 1.5    Benefits of Learning Epistemic Themes in Chemistry Education

The preceding arguments illustrate that much of the interest in the intersection of philosophy of chemistry and chemistry education has been expressed in theoretical form for the inclusion of philosophical ideas in chemistry education. The empirical examination of such arguments remains minimal particularly in relation to research on teacher education (e.g. Vesterinen et al. 2013). In this book, we aim to contribute to the debates on how philosophical ideas on nature of chemistry can be integrated meaningfully in chemistry teacher education. As such, we move beyond theoretical positions on how and why epistemic aspects of chemistry should be taught and learnt towards an educational empirical research that focuses on what is possible to accomplish in chemistry education in relation to epistemic content and context of chemistry. The applications of epistemological perspectives on chemistry in chemistry education can be accomplished in many ways. For example, there can be input into design of innovative chemistry curricula. Pedagogical strategies and learning resources can be developed to incorporate them in teaching and learning scenarios in chemistry lessons. While a multitude of efforts and contributions are needed for broader impact, our approach to this issue has been in the context of pre-service teacher education. We recognise the centrality of teachers in the enactment of any curricular and pedagogical innovation, and the point of pre-service teacher education seems appropriate to ensure that future teachers are well equipped from the beginning of their preparation for the teaching profession. Although there is evidence on the challenges and obstacles for the incorporation of HPS in science education (e.g. Höttecke & Silva, 2011), the question remains about how best to approach pre-service teachers' appropriation of epistemic understanding.

Ultimately, the inclusion of epistemic themes in chemistry education can potentially benefit students' learning in numerous ways. Epistemic themes by definition are meta-level characterisations. For example, an epistemic issue relevant for chemistry education is the idea that chemists have particular epistemic aims and values such as objectivity and accuracy. Awareness of these aims and values would make explicit in students' minds as to why chemists do what they do and what criteria drive their decision-making. A framework proposed by Tsai (2004) helps summarise some of the benefits for students' philosophical perspectives. Tsai refers to points in highlighting students' regulation of their learning: epistemological commitment, metacognition and critical thinking. He defines epistemological commitment as "one of the major features in an individual's conceptual ecology, the dynamic interaction among conceptual and cognitive domains" (Tsai, 2004, p. 970). Epistemological commitments involve an individual's explanatory ideals, views on specific ideas about what counts as a successful explanation in the field like chemistry and general views about the nature of knowledge. They are also evaluative standards to judge the merits of knowledge, such as its generalisability and internal consistency. Conceptual conflicts may occur between an individual's personal (sometimes scientifically naive) knowledge and the formal ideas of a discipline (Hewson & Hewson, 1984). Tsai gives the example where if a person believes that chemical laws do not have internal consistency, he or she may learn the laws by rote and in isolation without coherent understanding. Furthermore, the learner may not recognise a logical conflict if two contradictory scientific ideas exist in his or her cognitive structures.

Tsai's second theme concerns metacognition. Metacognition is broadly defined as the ability to reflect on one's actions and thoughts. White and Mitchell (1994) define metacognition as "knowledge of the processes of thinking and learning, awareness of one's own, and the management of them" (p. 27). In this sense, metacognition is a self-regulatory skill (or relevant knowledge) whereby the learner monitors his or her own learning processes. According to Gunstone (1991), "an appropriately metacognitive learner is one who can effectively undertake the constructivist processes of recognition, evaluation and, where needed, reconstruction of existing ideas" (pp. 135–136). The third theme discussed by Tsai (2004) is critical thinking. Broadly speaking, critical thinking is the capacity to apply logical processes to scrutinise evidence. Critical thinking also has been defined in various ways by different researchers. For example, Lipman (1988) defines critical thinking as "skilful, responsible thinking that facilitates good judgment because it (a) relies upon criteria, (b) is self-correcting and (c) is sensitive to context" (p. 39). Tsai argues that the process of deciding what to believe or do depends on the learner's epistemological commitments (i.e. his or her standards of judging knowledge), and the use of reflective thinking depends on his or her metacognitive processing. In this sense, critical thinking is a metacognitive, rather than a cognitive skill (Kuhn, 1999).

Immersing students in epistemologically rich educational contexts is likely to facilitate the development of their epistemological commitments, metacognition

**Table 1.1**  Potential benefits of learning the epistemic core of chemistry

| Aspect of epistemic core | Epistemological commitments | Metacognition | Critical thinking |
|---|---|---|---|
| Aims and values | Students appropriate a set of epistemic aims and values from chemistry such as commitment to accurate and objective evidence | Students are aware of their use of epistemic aims and values of chemistry in their investigations | Students can evaluate whether or not chemists' claims are in line with the epistemic aims and values of chemistry |
| Practices | Students are committed to employing appropriate chemical practices such as modelling and classification in investigating problems | Students can evaluate their understanding of chemical practices such as modelling and classification | Students can compare and contrast the strengths and limitations of different practices such as experimentation and observation |
| Methods | Students value the importance of diversity of methods in chemistry ranging from hypothesis testing to non-manipulative observation | Students can distinguish between different methods in chemistry and select them to be fit for purpose in problem-solving | Students can advance arguments for and against the use of a particular method to investigate a problem |
| Knowledge | Students understand that chemistry relies on different forms of knowledge such as theories, laws and models and that these knowledge forms develop in time | Students can evaluate their own chemistry knowledge and characterise it relative to established theories, models and laws in chemistry | Students understand the explanatory power of chemistry knowledge as well as its limitations |

and critical thinking. In Chap. 2, we will introduce the idea of the "epistemic core" to illustrate how particular epistemic themes can be framed for educational purposes. The framework is inspired by the work of Erduran and Dagher (2014a) which included a broader characterisation of the cognitive-epistemic systems of science. The "epistemic core" as we define it signifies the centrality of certain epistemic categories and it is related to the aims and values, practices, methods and knowledge in science. As such, the "epistemic core" is about how and why scientific knowledge is constructed, evaluated and revised. Although the details of these categories will be unpacked subsequently in Chap. 2, it is worthwhile to explore the potential benefits for students' learning of epistemological perspectives more broadly and in relation to the epistemic core in particular. Table 1.1 provides some example contributions of the learning of the epistemic core of chemistry to other learning outcomes such as epistemological commitments, metacognition and critical thinking. For example, in the case of methods, students' critical thinking skills would be improved if they could produce arguments for and against the use of a particular method to investigate a problem (e.g. filtration to separate insoluble sand from sand-water mixture versus evaporation to separate salt dissolved in water). Here the students would not only consider the particular details of the methods but also reflect on the methods at a more meta-level.

## 1.6    Rationale and Outline of the Book

Although Table 1.1 provides some examples and specifications about how the epistemic core could relate to potential student learning outcomes, the ideas represented are fairly abstract. From a pedagogical point of view, for example, it is not entirely clear from such a specification how students or teachers, for that matter, could be supported in understanding diversity of methods at a meta-level so that they can resort to choosing particular methods if a particular scenario calls for them. Given that the epistemic aspects of chemistry are fairly unfamiliar to teachers and students alike, pedagogical tools are essential to facilitate their teaching and learning. Several strategies exist to facilitate the teaching and learning of the epistemic core. First, visualisation can be used as a way to summarise and represent some fairly complex ideas. Erduran and Dagher (2014a) proposed what they called the "Generative Images of Science". These are visual representations that capture some key discussion points, and they are derived from a review of philosophy of science literature. For example, their Benzene Ring Heuristic captures some rather complex ideas about how scientific practices such as modelling and predictions are mediated through representations and discourse. The *Generative Images of Science* are thus meant to distill some complex ideas for visually representing them for educators' purposes. In other words, they were already informed by design with educators in mind. Hence, we capitalise on Erduran and Dagher's (2014a) *Generative Images of Science* which were derived from a review of philosophy of science literature and proposed for pragmatic educational purposes.

The overall aims of this book are twofold: (a) to synthesise theoretical perspectives using philosophy of chemistry and teacher education, traditionally disparate bodies of literature, and (b) to explore how such perspectives can be infused into the design and implementation of teacher education programmes to impact pre-service teachers. While countless arguments have been made about the integration of philosophy of science in science teacher education more broadly (e.g. Matthews, 2014), there is paucity of empirical evidence on what content can be relevant and appropriate for teacher education and what is possible to accomplish with pre-service teachers given their limited exposure to epistemological themes. It is important to note that while philosophy of chemistry is the foundational rationale of the book in relation to why epistemic aspects of chemistry are important, the research on direct input of philosophy of chemistry in chemistry teacher education is in its infancy. Given how unfamiliar philosophical perspectives are in chemistry education, there is an elementary task of introducing the relevance of epistemic themes in chemistry pre-service teacher education.

Hence, while the theoretical work from philosophy of chemistry might provide us with clues as to what to teach, it does not point to how it should be taught or indeed what is teachable content for teachers from different career stages. Similarly, shifting the focus from the concepts and processes of chemistry to epistemic thinking might seem as if the chemistry content is being dumbed down. While the destination of chemistry education may be the full integration of epistemic ideas that are situated in and coupled with chemistry subject knowledge, as the book will illustrate, the process of getting there is filled with various constraints that need to be taken seriously and addressed directly if the objective is to have a real impact on the practice of chemistry education. While theoretical arguments may be acceptable for what particular philosophical themes are important in chemistry and therefore worthy of teaching in chemistry, their incorporation in chemistry education practice, including teacher education, may present significant challenges ranging from the limited background of the learners as well as the teachers to institutional barriers. For example, it is unlikely that pre-service chemistry teachers would have had any formal exposure to philosophy of science as part of their academic background. Indeed, many chemistry teacher educators' themselves are unlikely to have had formal training in philosophy of science. Hence, the content of what aspects and content of philosophy of chemistry can be infused in teacher education has to be considered realistically relative to what is feasible to be accomplished. Our intention in this book is not to provide some conclusive endpoint recommendations for how to infuse philosophy of chemistry in chemistry teacher education. Rather, it is to illustrate the beginning point of what is feasible to accomplish within pre-service teacher education with participants and teacher educators who have limited background on the subject. As such, the primary goal is to motivate epistemic thinking at a basic level with potential for progression to higher levels of epistemic thinking situated in robust chemistry examples. In this book, we are emphasising pre-service teachers' learning of epistemic themes because this is the primary objective of the book. However, we do recognise that historical context of such themes is important as well. The area of work can serve as a starting point for a research programme that explores trajectories in the development of pre-service teachers, teacher educators and teacher preparation programmes that eventually consolidate not only the epistemic themes but also the historical context of chemistry concepts.

As noted, there will be a focus on the *epistemic core* which includes the aims and values, practices, methods and knowledge in science. Erduran and Dagher (2014a) produced images for each category. The epistemic core is a domain-general (e.g. Kampourakis, 2016) characterisation about science at large. These categories signify the knowledge production, evaluation and revision processes in science. However, each category of the epistemic core can be instantiated in a specific domain such as chemistry. In Chap. 2, we illustrate how the epistemic core and the related visual representations from the set of *Generative Images of Science* can be related to the domain of chemistry. The articulation of what epistemic ideas should be included in chemistry education is one thing; the actual implementation of these ideas in practice is another. When teachers themselves have not gone through an education system that empowers them to understand the epistemic undertones of the knowledge that they are teaching in lessons, it is unreasonable to expect that they will be skilled to teach about these abstract ideas.

A major duty rests upon teacher education to enhance current and future chemistry teachers' understanding of epistemic themes and their skills in teaching them in meaningful ways that are appropriate for the cognitive levels of the students. We present in-depth examples of the epistemic core as a theme to be covered in teacher education. For each category, we examine how they can be situated in chemistry examples and explore the implications for chemistry education. For example, we will illustrate an example of an epistemic aim in chemistry with "simplicity". Hoffman, Minkin, and Carpenter (1997) discuss the application of this aim in the context of reaction mechanisms in chemistry. They give the example that a definition of reaction mechanism can involve an obvious analogy with the mechanical description of particles. They refer to the work of Rice and Teller (1936) who proposed the "principle of least motion" (PLM) according to which "those elementary reactions will be favored that involve the least change in atomic position and electronic configuration" (p. 489). The particular instances of the epistemic core situated in chemistry subject knowledge provide a kind of destination as content for teacher education programmes. Although these are meant to be the eventual learning outcomes for teachers of chemistry, the process of educating future teachers to get there involves a much more elementary starting point. This is because the background of future teachers in epistemic aspects of their subjects is typically very limited. Hence they need to be introduced to why epistemic thinking in chemistry is important and why it should therefore be a component of chemistry education.

The goal of teacher education is bound to be unfulfilled unless it considers empirical evidence on how real teachers think and how they learn. Hence in Chap. 3, we turn our attention to the research literature on teachers' beliefs, learning and knowledge to investigate the evidence on how best to consider pre-service teachers' learning of the epistemic core. Research on teachers' beliefs, knowledge and learning provide a context for the inclusion of epistemic core in teacher education. We illustrate the relevance and link of the "epistemic core" framework for research and development in pre-service teacher education. We relate the idea of the "epistemic core" to some aspects of teachers' learning including teachers' meta-strategic knowledge (Zohar, 2012) and epistemic cognition (Greene, Sandoval, & Braten, 2016). The reference to research on teacher education sets the foundation for realistic recommendations for improving teacher education. Furthermore, a survey of teacher education provision internationally highlights some programme and state level constraints including the high degree of accountability. As our own institutional contexts in England and Turkey also illustrate, there can be significant variations in how teacher education programmes are structured in different countries. For example, the duration of pre-service teacher education programmes can range from a few weeks to almost 2 years in the case of certification programmes and 4–5 years in undergraduate programmes. Given the vast differences in our own professional contexts, a brief contrast of pre-service teacher education provision at our own universities provides examples to illustrate the challenges as well as opportunities for incorporating epistemic themes.

In Chap. 4, we turn to the design and implementation of a teacher education intervention that includes sets of resources and strategies derived empirically. Particular strategies such as argumentation (e.g. Erduran & Jimenez-Aleixandre, 2007), visualisation (e.g. Eilam & Gilbert, 2014) and analogies (e.g. Aubusson, Treagust, & Harrison, 2009) have been promoted in the teacher education context to be reported in subsequent chapters. The emphasis in this chapter is the articulation of the design features of teacher education programmes that can help support the teaching of epistemic aspects of chemistry. A module designed to incorporate pre-service teachers' active learning of aims and values, practices, methods and knowledge in science are described. As previously stated, the epistemic core idea is derived from Erduran and Dagher's (2014b) work which was based on an approach to nature of science called the "Family Resemblance Approach" (FRA) (e.g. Irzik & Nola, 2011, 2014). There is now a growing number of studies using FRA as a framework in science education (e.g. Akgun, 2018; Alayoglu, 2018; BouJaoude, Dagher & Refai, 2017; Cullinane, 2018; Dagher & Erduran, 2016, 2017; Erduran, 2014; 2017; Erduran & Kaya, 2018; Karabas, 2017; Kaya & Erduran, 2016; Kaya, Erduran, Aksoz & Akgun, 2019; McDonald, 2017). FRA essentially describes sciences as part of a family of related disciplines which share particular features such as aims, values and methods, while at the same time they may also differ in their nuanced approaches to each aspect. This foundational issue about how to justify a science as "science" is also incorporated into the design of the teacher education intervention. The context for teacher education was an undergraduate module taught by ourselves as the authors of the book at a state university in Turkey. The pre-service teachers were senior year students in a 5-year teacher preparation programme. The teacher education intervention lasted for 11 3 h sessions. The chapter discusses in detail the content of each session including the tasks and strategies used, and it provides examples of ideas produced by pre-service teachers.

In Chap. 5, we show how the goals and content in the intervention had direct and close impact on pre-service teachers by illustrating a close link between the "input and output" of key themes. The chapter is organised around these themes (e.g. diversity of methods, growth of knowledge) that formed the basis of the design and implementation of the teacher education intervention. Data from sessions and interviews with pre-service teachers are used to illustrate the influence of the teacher education intervention on three pre-service chemistry teachers. The in-depth analysis of these teachers' outputs illustrates how they interpreted the epistemic themes presented through the epistemic core. In Chap. 6, we exemplify the case of a struggling pre-service teacher, Alev, to highlight what is possible to accomplish in pre-service teachers' learning of the key epistemic core ideas when there is limited academic skills more generally. By focusing on her representations and perceptions before and after the teacher education intervention, we present a "thick description" of the impact of the module. The overall pattern of impact on Alev is illustrated through the particular emphases in the intervention (e.g. diversity of methods, growth of knowledge). Both Chaps. 5 and 6 show how the teacher education intervention included and resulted in analogical reasoning and visualisation about the epistemic core.

Following on the discussion on the design and implementation of the teacher education intervention, we turn to ourselves as teacher educators to report on a self-study in Chap. 7. The underpinning assumption in this chapter is that understanding the identity of teacher educators is critical to ensuring that teacher education can be innovative and progressive (e.g. Swennen, Jones, & Volman, 2010) in incorporating new and challenging content for inclusion in the training of future teachers. Teacher educators' identities can be complex, as our own examples illustrate. Being teacher educators in higher education often involves working in different contexts, such as schools, higher education administration and research (Ellis, Blake, McNicholl, & McNally, 2011). Hence, they may have variations in their identities which can be multifaceted.

We are motivated to pursue a reflective account of our own experiences because (a) the broad content of the book, the interplay between philosophy of chemistry and chemistry education, is fairly marginal and some of our reflections could potentially benefit others as they take on a similar approach in their work, (b) our reflections can potentially serve as data sources that help interpret the general project of infusing epistemic aspects of chemistry in chemistry teacher education and (c) our reflections may provide colleagues with realisations that might help improve their practices as teacher educators and chemistry education researchers. We present our reflections including our journeys into teacher education, background in history and philosophy of science and views on issues related to the incorporation of nature of chemistry in teacher education more broadly. Through this reflective account, we have come to re-envisage ourselves as teacher educators who are trying to make sense of some rather abstract and deep philosophical ideas ourselves. We have had to negotiate our professional identities through differentiated expertise in relation to science education research, philosophy of chemistry and teacher education, thus questioning our own professional knowledge.

In finalising the book in Chap. 8, we consolidate our discussion so far by highlighting some of the limitations as well as the contributions to research and development on epistemological themes in chemistry education. Contextualising the chapters in the book in the newly emerging science identity research literature (e.g. Avraamidou, 2014), we propose a framework on the development of "epistemic identity" in pre-service teacher education and continuous professional development. The framework provides a coordinated and systemic approach to addressing the fundamental question of how future and existing teachers of chemistry can be supported to become inquisitive about what knowledge in chemistry is about as well as how it is constructed and justified. The book concludes with some suggestions for future research.

## 1.7   Conclusions

Arguments in chemistry education for the inclusion of epistemic aspects of chemistry need to be mindful of existing research on learning and pedagogy. It is one thing to suggest that some philosophical ideas are potentially important for chemistry education, and another to frame such proposals with knowledge about empirical educational evidence on how teachers think and how they are prepared to teach. Our decision-making in this book in relation to the selection of epistemic themes has been based on our experiences as chemistry teacher educators and chemistry education researchers. The major task of this book is not one of direct import of ideas from philosophy of chemistry. It is to ensure that such ideas can be usable, intelligible and relevant for chemical education research and practice. The selection of content from philosophy of chemistry for teacher education purposes demands consistency with other goals including pedagogical goals and curriculum context. Teachers do not operate within a vacuum of interesting ideas. They are expected to teach to a particular curriculum or a programme of work where there is a high degree of accountability in terms of high-stakes assessments of their students. Hence, even when some ideas from philosophy of chemistry might be cognitively appropriate for students, the constraints that teachers, and teacher educators for that matter, need to be acknowledged and worked with if there will be any incorporation of epistemic themes in chemistry education at all.

Key questions such as "What is chemical knowledge and how does it develop? What criteria, standards and heuristics shape its development?" are directly relevant for ensuring that teaching and learning environments are effectively structured and resourced for sound and deep understanding of chemistry (Erduran, 2009). While theoretical investigations can help orient the design of educational content ranging from curriculum to teacher education informed by significant philosophical issues in chemistry, the actual implementation of such curricula demand more than rhetoric. Teachers' role in the implementation of meta-perspectives on chemistry is crucial, and teachers need to be supported in understanding unfamiliar ideas. In this sense, a great responsibility rests on teacher educators in developing teachers' understanding of the epistemic aspects of chemistry. Some crucial questions are thus raised for teacher education: *What strategies can be used to facilitate preservice teachers understanding of epistemic themes in chemistry? How can teacher educators monitor pre-service teachers' understanding of the epistemic core so that they can provide formative feedback? Who are the teacher educators engaged in the teaching of the epistemic core to pre-service teachers and how can their own professional development be enhanced for their inclusion of epistemic themes in their teaching?*

As a final note of reflection about the interactions of philosophy of chemistry and chemistry education, the complexity associated with the foundations of chemistry education should be noted. Chemistry education is not only about chemistry. It is also about education. Education by definition is a multidisciplinary field, drawing on perspectives from disciplines such as cognitive psychology, sociology and philosophy of education and policy studies. Arguments surrounding empirical research on education often appeal to policy and practice to frame the actual operation of education systems in particular contexts. The goals and aims of education do not necessarily correspond to the goals and aims of chemistry research nor philosophical analysis. Educational researchers have to draw ideas from a multitude of perspectives such as philosophy of chemistry, but they also need to converge on empirically sound and educationally realistic and pragmatic agendas. Our intention is therefore to explore the feasibility of transforming some highly abstract philosophical ideas into practical pedagogical contexts, thereby contributing to a much-needed evidence base on the utility of philosophy of chemistry for chemistry education.

# References

Aduriz-Bravo, A. (2013). A 'semantic' view of scientific models for science education. *Science & Education, 22*(7), 1593–1611.

Akgun, S. (2018). *University students' understanding of the nature of science.* Unpublished Master's thesis. Bogazici University, Istanbul, Turkey.

Alayoglu, M. (2018). *Fifth-grade students' attitudes towards science and their understanding of its social-institutional aspects.* Unpublished Master's thesis. Bogazici University, Istanbul, Turkey.

Aubusson, P., Treagust, D., & Harrison A. (2009). Learning and teaching science with analogies and metaphors. In *The world of science education: Handbook of research in Australasia.* Rotterdam, The Netherlands: Sense Publishers.

Avraamidou, L. (2014). Studying science teacher identity: Current insights and future research directions. *Studies in Science Education, 50*(2), 145–179.

Baird, D., Scerri, E., & McIntyre, L. (2006). Introduction: The invisibility of chemistry. In D. Baird, E. Scerri, & L. McIntyre (Eds.), *Philosophy of chemistry: Synthesis of a new discipline.* Dordrecht, The Netherlands: Springer.

BouJaoude, S., Dagher, Z., & Refai, S. (2017). The portrayal of nature of science in Lebanese ninth grade science textbooks. In C. V. McDonald & F. Abd-el-Khalick (Eds.), *Representations of nature of science in school science textbooks: A global perspective* (pp. 79–97). New York: Routledge.

Chamizo, J. A. (1992). La química en secundaria, o por qué la enseñanza moderna de la química no es la enseñanza de la química moderna [Chemistry in junior high school, or why modern teaching of chemistry is not modern chemistry teaching]. *Información Científica y Tecnológica, 14,* 49–51.

Chamizo, J. A. (2013). A new definition of models and modeling in chemistry's teaching. *Science & Education, 22*(7), 1613–1632.

Chamizo, J. A. (2014). The role of instruments in three chemical' revolutions. *Science Education, 23,* 955–982.

Coenders, S. F., Terlouw, C., Dijkstra, S., & Pieters, J. (2010). The effects of the design and development of a chemistry curriculum reform on teachers' professional growth: A case study. *Journal of Science Teacher Education, 21,* 535–557.

Conant, J. B. (1947). *On understanding science.* New Haven, CT: Yale University Press.

Conant, J. B. (1948). *Harvard case histories of experimental science.* Cambridge, MA: Harvard University Press.

Cooper, M. (2018). Chemistry education research- from personal empiricism to evidence, theory and informed practice. *Chemical Reviews, 118*(12), 6053–6087.

Cullinane, A. (2018). *Incorporating nature of science in initial science teacher education.* Unpublished PhD dissertation. University of Limerick, Ireland.

Dagher, Z., & Erduran, S. (2016). Reconceptualizing the nature of science: Why does it matter? *Science & Education, 25*(1 & 2), 147–164.

Dagher, Z., & Erduran, S. (2017). Abandoning patchwork approaches to nature of science in science education. *Canadian Journal of Science, Mathematics and Technology Education, 17*(1), 46–52.

Duschl, R. (1990). *Restructuring science education: The importance of theories and their development.* New York: Teachers College Press.

Duschl, R., Erduran, S., Grandy, R., & Rudolph, J. (2006). Guest editorial: Science studies and science education. *Science Education, 90*(6), 961–964.

Early, J. E. (2013). A new 'idea of nature' for chemical education. *Science & Education, 22*(7), 1775–1786.

Eilam, B., & Gilbert, J. K. (2014). *Science teachers' use of visual representations.* Dordrecht, The Netherlands: Springer.

Ellis, V., Blake, A., McNicholl, J., & McNally, J. (2011). The work of teacher education: The final research report for the *Higher Education Academy, Subject Centre for Education*. ESCalate.

Erduran, S. (2001). Philosophy of chemistry: An emerging field with implications for chemical education. *Science & Education, 10*, 581–593.

Erduran, S. (2005). Applying the philosophical concept of reduction to the chemistry of water: Implications for chemical education. *Science & Education, 14*(2), 161–171.

Erduran, S. (2007). Breaking the law: Promoting domain-specificity in science education in the context of arguing about the periodic law in chemistry. *Foundations of Chemistry, 9*(3), 247–263.

Erduran, S. (2009). Beyond philosophical confusion: Establishing the role of philosophy of chemistry in chemical education research. *Journal of Baltic Science Education, 8*(10), 5–14.

Erduran, S. (2013). Editorial: Philosophy, chemistry and education: An introduction. *Science & Education, 22*, 1559.

Erduran, S. (2014). Beyond nature of science: The case for reconceptualising 'science' for science education. *Science Education International, 25*(1), 93–111.

Erduran, S. (2017). Visualising the nature of science: Beyond textual pieces to holistic images in science education. In K. Hahl, K. Juuti, J. Lampiselkä, J. Lavonen, & A. Uitto (Eds.), *Cognitive and affective aspects in science education research: Selected papers from the ESERA 2015 conference* (pp. 15–30). Dordrecht, The Netherlands: Springer.

Erduran, S., & Dagher, Z. (2014a). *Reconceptualizing the nature of science for science education: Scientific knowledge, practices and other family categories*. Dordrecht, The Netherlands: Springer.

Erduran, S., & Dagher, Z. (2014b). Regaining focus in Irish junior cycle science: Potential new directions for curriculum development on nature of science. *Irish Educational Studies, 33*(4), 335–350.

Erduran, S., & Jimenez-Aleixandre, M. P. (Eds.). (2007). *Argumentation in science education: Perspectives from classroom-based research*. Dordrecht, The Netherlands: Springer.

Erduran, S., & Kaya, E. (2018). Drawing nature of science in pre-service science teacher education: Epistemic insight through visual representations. *Research in Science Education, 48*(6), 1133–1149.

Fernandez-Gonzalez, M. (2013). Idealization in chemistry: Pure substance and laboratory product. *Science & Education, 22*(7), 1723–1740.

Gabel, D., & Bunce, D. (1984). Research on problem solving in chemistry. In D. Gabel (Ed.), *Handbook of research on science teaching and learning* (pp. 301–326). New York: Macmillan Publishing Company.

Garritz, A. (2013). Teaching the philosophical interpretations of quantum mechanics and quantum chemistry through controversies. *Science & Education, 22*(7), 1787–1807.

Gilbert, J., & Boulter, C. (2000). *Developing models in science education*. Dordrecht, The Netherlands: Kluwer Academic.

Gilbert, J. K. (2006). On the nature of "context" in chemical education. *International Journal of Science Education, 28*, 957–976.

Gilbert, J. K., de Jong, O., Justi, R., Treagust, D. F., & van Driel, J. H. (2003). Research and development for the future of chemical education. In J. K. Gilbert, O. De Jong, R. Justi, D. F. Tragust, & J. H. van Driel (Eds.), *Chemical education: Towards research based practice* (pp. 391–408). Dordrecht, The Netherlands: Kluwer.

Good, R. J. (1999). Why are chemists turned off by philosophy? *Foundations of Chemistry, 1*, 65–96.

Greene, J. A., Sandoval, W. A., & Braten, I. (2016). *Handbook of epistemic cognition*. New York: Routledge.

Gunstone, R. F. (1991). Constructivism and metacognition: Theoretical issues and classroom studies. In R. Duit, F. Goldberg, & H. Niedderer (Eds.), *Research in physics learning: Theoretical issues and empirical studies* (pp. 129–140). Kiel, Germany: Institute of Science Education.

Hewson, P. W., & Hewson, M. G. (1984). The role of conceptual conflict in conceptual change and the design of science instruction. *Instructional Science, 13*(1), 1–13.

Hodson, D. (1988). Towards a philosophically more valid science curriculum. *Science Education, 72*, 19–40.

Hodson, D. (2011). *Looking to the future: Building a curriculum for social activism.* Rotterdam, The Netherlands: Sense Publishers.

Hoffman, R., Minkin, V. I., & Carpenter, B. K. (1997). Ockham's razor and chemistry. *Hyle – An International Journal for the Philosophy of Chemistry, 3*, 3–28.

Höttecke, D., & Silva, C. C. (2011). Why implementing history and philosophy in school science education is a challenge: An analysis of obstacles. *Science & Education, 20*, 293–316.

Irzik, G., & Nola, R. (2011). A family resemblance approach to the nature of science for science education. *Science & Education, 20*, 591–607.

Irzik, G., & Nola, R. (2014). New directions for nature of science research. In M. Matthews (Ed.), *International handbook of research in history, philosophy and science teaching* (pp. 999–1021). Dordrecht, The Netherlands: Springer.

Izquierdo-Aymerich, M. (2013). School chemistry: An historical and philosophical approach. *Science & Education, 22*(7), 1633–1653.

Johnstone, A. H. (1993). The development of chemistry teaching: A changing response to changing demand. *Journal of Chemical Education, 70*, 701–705.

Kampourakis, K. (2016). The "general aspects" conceptualization as a pragmatic and effective means to introducing students to nature of science. *Journal of Research in Science Teaching, 53*(5), 667–682.

Karabas, N. (2017). *The effect of scientific practice-based instruction on seventh graders' perceptions of scientific practices.* Unpublished Master's thesis. Bogazici University, Istanbul, Turkey.

Kauffman, G. B. (1989). History in the chemistry curriculum. *Interchange, 20*(2), 81–94.

Kaya, E., & Erduran, S. (2013). Integrating epistemological perspectives on chemistry in chemical education: The cases of concept duality, chemical language, and structural explanations. *Science & Education, 22*(7), 1741–1755.

Kaya, E., & Erduran, S. (2016). From FRA to RFN, or how the family resemblance approach can be transformed for science curriculum analysis on nature of science. *Science & Education, 25*(9), 1115–1133.

Kaya, E., Erduran, S., Aksoz, B., & Akgun, S. (2019). Reconceptualised family resemblance approach to nature of science in pre-service science teacher education. *International Journal of Science Education, 41*(1), 21–47.

Kelly, G. J. (2011). Scientific literacy, discourse, and epistemic practices. In C. Linder, L. Östman, D. A. Roberts, P. Wickman, G. Erikson, & A. McKinnon (Eds.), *Exploring the landscape of scientific literacy* (pp. 61–73). New York: Routledge.

Klopfer, L. (1969). The teaching of science and the history of science. *Journal of Research in Science Teaching, 6*, 87–95.

Kuhn, D. (1999). A development model of critical thinking. *Educational Researcher, 28*(2), 16–46.

Laszlo, P. (1999). Circulation of concepts. *Foundations of Chemistry, 1*, 225–238.

Laszlo, P. (2013). Towards teaching chemistry as a language. *Science & Education, 22*(7), 1669–1706.

Lipman, M. (1988). Critical thinking – what can it be? *Educational Leadership, 46*(1), 38–43.

Lythcott, J. (1990). Problem solving and requisite knowledge of chemistry. *Journal of Chemical Education, 67*(3), 248–252.

Mahaffy, P. (2006). Moving chemistry education into 3D: A tetrahedral metaphor for understanding chemistry. *Journal of Chemical Education, 83*(1), 49–55.

Matthews, M. R. (2014). *Science teaching. The role of history and philosophy of science.* New York: Routledge.

McDonald, C. V. (2017). Exploring representations of nature of science in Australian junior secondary school science textbooks: A case study of genetics. In C. V. McDonald & F. Abd-el-Khalick (Eds.), *Representations of nature of science in school science textbooks: A global perspective* (pp. 98–117). New York: Routledge.

McIntyre, L., & Scerri, E. (1997). The philosophy of chemistry- editorial introduction. *Synthese, 111*(3), 211–212.

Newman, M. (2013). Emergence, supervenience, and introductory chemical education. *Science & Education, 22*(7), 1655–1667.

Niaz, M. (1988). Manipulation of M demand of chemistry problems and its effect on student performance: A neo-Piagetian study. *Journal of Research in Science Teaching, 25*, 643–657. https://doi.org/10.1002/tea.3660250804.

Niaz, M. (2011). *Innovating science teacher education: A history and philosophy of science perspective*. Oxon, UK: Routledge.

Niaz, M. (2016). *Chemistry education and contributions from history and philosophy of science*. Dordrecht, The Netherlands: Springer.

Pedretti, E., & Nazir, J. (2011). Currents in STSE education: Mapping a complex field. *Science Education, 95*, 601–626.

Pinto-Ribeiro, M. A., & Costa-Pereira, D. (2013). Constitutive pluralism of chemistry: Thought planning, curriculum, epistemological and didactic orientations. *Science & Education, 22*(7), 1809–1837.

Rice, F., & Teller, E. (1936). The role of free radicals in elementary organic reactions. *Journal of Chemical Physics, 6*, 489.

Ross, B., & Munby, H. (1991). Concept mapping and misconceptions: A study of high-school students' understanding of acids and bases. *International Journal of Science Education, 13*(1), 11–23.

Sandoval, W. (2005). Understanding students' practical epistemologies and their influence on learning through inquiry. *Science Education, 89*, 634–656.

Scerri, E. (1997). Are chemistry and philosophy miscible? *The Chemical Intelligencer, 3*, 44–46.

Scerri, E. (2000). Philosophy of chemistry-A new interdisciplinary field? *Journal of Chemical Education, 77*(4), 522–525.

Scerri, E. R. (2007). *The periodic table: Its story and its significance*. New York: Oxford University Press.

Sjöström, J. (2013). Towards Bildung-oriented chemistry education. *Science & Education, 22*(7), 1873–1890.

Standen, A. (1948). Three ways of teaching chemistry. *Journal of Chemical Education, 25*(9), 506.

Swennen, A., Jones, K., & Volman, M. (2010). Teacher educators: Their identities, sub-identities and implications for professional development. *Professional Development in Education, 36*(1–2), 131–114.

Talanquer, V. (2013). School chemistry: The need for transgression. *Science & Education, 22*(7), 1757–1773.

Teo, T. W., Goh, M. T., & Yeo, L. W. (2014). Chemistry education research trends: 2004–2013. *Chemistry Education Research and Practice, 15*, 470–487.

Thalos, M. (2013). The lens of chemistry. *Science & Education, 22*(7), 1707–1721.

Tobin, E. (2013). Chemical laws, idealization and approximation. *Science & Education, 22*(7), 1581–1592.

Tsai, C.-C. (2004). A review and discussion of epistemological commitments, metacognition, and critical thinking with suggestions on their enhancement in internet-assisted chemistry classrooms. *Journal of Chemical Education, 78*(7), 970–974.

van Brakel, J. (1997). Chemistry as the science of the transformation of substances. *Synthese, 111*(3), 253–282.

van Brakel, J. (2000). *Philosophy of chemistry: Between the manifest and the scientific image*. Leuven, Belgium: Leuven University Press.

van Brakel, J. (2010). A subject to think about: Essays on the history and philosophy of chemistry. *Ambix, 57*(2), 233–234.

Vesterinen, V. M., Aksela, M., & Lavonen, J. (2013). Quantitative analysis of representations of nature of science in Nordic secondary school textbooks using framework of analysis based on philosophy of chemistry. *Science & Education, 22*(7), 1839–1855.

Vilches, A., & Gil-Perez, D. (2013). Creating a sustainable future: Some philosophical and educational considerations for chemistry teaching. *Science & Education, 22*(7), 1857–1872.

Weisberg, M., Needham, P., Hendry, R. (2011). Philosophy of chemistry. In *Stanford encyclopedia of philosophy*. Stanford, CA: Stanford University.

White, R. T., & Mitchell, I. J. (1994). Metacognition and the quality of learning. *Studies in Science Education, 23*, 21–37.

Woody, A. (2013). How is the ideal gas law explanatory? *Science & Education, 22*(7), 1563–1580.

Zohar, A. (2012). Explicit teaching of metastrategic knowledge: Definitions, students' learning, and teachers' professional development. In A. Zohar & Y. J. Dori (Eds.), *Metacognition in science education: Trends in current research* (Vol. 40, pp. 197–223). Dordrecht, The Netherlands: Springer.

# Chapter 2
# Defining the Epistemic Core of Chemistry

## 2.1 Introduction

Chemistry is a domain that is concerned with the analysis and synthesis of substances. Chemists are interested in understanding the structure and function of matter. They engage in particular epistemic processes that are underpinned by particular aims, values and practices for the purpose of generating reliable knowledge to explain the material world. The process of knowledge generation involves debate about data as data get classified and reclassified and models are constructed to account for observed properties. For example, "organic compounds were initially defined according to their principal source, animal or vegetable, and according to their main chemical functions, acid, base, fat, dye and so on" (Brock, 1992, p.211). Accumulation of knowledge about organic compounds, however, led to reconsideration suggesting that it might be possible to classify organic compounds based on chemical criteria. Berzelius proposed the electrochemical theory which emphasised that organic substances obey the same laws as inorganic substances (Leicester, 1981). He argued that organic compounds always consisted of oxygen combined with a compound radical. Plant substances in this view consisted of carbon and hydrogen while animal substances consisted of carbon, hydrogen and nitrogen. It was not until much uncertainty and a period of further investigations that Berzelius proposed isomerism after failing to detect any compositional difference between racemic and tartaric acids, leading to one of the fundamental ideas of organic chemistry: the abundance of carbon in organic compounds.

History and philosophy of chemistry as well as contemporary chemistry are rich with such examples that illustrate how chemistry is guided by a set of epistemic aims (Bhushan & Rosenfeld, 2000). Chemistry has had a strong empirical dimension, and the empirical adequacy of proposed explanations needs to be justified. As the Berzelius example illustrates, chemists sometimes engage in taxonomic classification of substances that rely on observations and hypothesis testing but may not

© Springer Nature Switzerland AG 2019
S. Erduran, E. Kaya, *Transforming Teacher Education Through the Epistemic Core of Chemistry*, Science: Philosophy, History and Education,
https://doi.org/10.1007/978-3-030-15326-7_2

necessarily involve manipulation of variables per se. The ultimate purpose of chemistry, of course, is to propose theories, laws and models that can explain observations and help make predictions. In this chapter, we propose a framework derived from the work of Erduran and Dagher (2014) to illustrate how an "epistemic core" of chemistry can be conceptualised in a way that can help inform school chemistry.

Epistemic aims, values, practices and methods of chemistry and chemistry knowledge are described. Erduran and Dagher's (2014) *Generative Images of Science* can serve as both domain-general and domain-specific tools (see Chap.1). By design, they are domain general, pointing to particular epistemic features of science. However they can be unpacked and articulated in any scientific context including chemistry. The chapter applies Erduran and Dagher's categories in the context of chemistry by reviewing the research literature on philosophy of chemistry. As such, the approach not only builds on these categories but also provides a purposeful synthesis for chemistry education. Erduran and Dagher's work was informed by broad themes and examples from philosophy of science. It did not exclusively include chemistry content because it did not aim to do so. Hence, the chapter's contribution is to illustrate chemistry examples and demonstrate further the domain-specific utility of their framework. It should be noted that Erduran and Dagher referred to "cognitive-epistemic" categories because they authors wanted to highlight that cognitive and epistemic themes are interrelated (e.g. Corlett, 1991). For the sake of simplicity and because of the emphasis on philosophy of chemistry in this book, we emphasise the epistemic aspect of the categories.

Considering the vast amount of literature that is now available in philosophy of chemistry (see Chap. 1), our intention is not to be exhaustive about epistemic aspects of chemistry. As chemistry educators, we are more concerned about using evidence from the philosophy of chemistry literature to inform chemistry education such that the teaching and learning of chemistry can be situated in more authentic contexts. The goal is to provide a coherent epistemic framework that can serve as a toolkit for educational purposes including the purposes of research and the design of teacher education interventions. The reference to philosophy of chemistry thus is guided by a pragmatic sense of utility and relevance for school chemistry. Furthermore, the content drawn from philosophy of chemistry needs to have some relevance to the chemistry curriculum, and it needs to be transformed into a form that is appropriate for the cognitive demands placed on teachers and students.

## 2.2  Aims and Values in Chemistry

Chemistry, like all of science, is guided by a set of aims and values that can be broadly categorised as cognitive, epistemic and social (Erduran & Dagher, 2014) (see Fig. 2.1). Given the theme of the book, we focus on the epistemic aims and values such as "objectivity" and "accuracy". Shapin and Shaffer (1985) illustrate

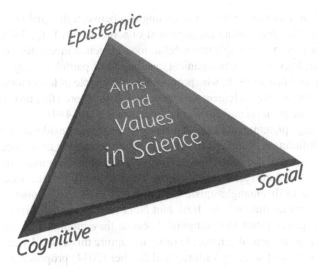

**Fig. 2.1** Aims and values of science. (From Erduran & Dagher, 2014, p. 49)

the importance of "objectivity" as an aim in chemistry by referring to Robert Boyle and how he sought to "let the air-pump speak" for itself rather than being clouded in human judgement. According to Allchin (1999), science and values intersect in at least three ways. First, there are epistemic values that guide research. For example, accuracy, testability and novelty can guide scientists in making judgements about knowledge claims. Second, science is situated in a particular cultural context, and thus it is inherently composed of and surrounded by values that are practised by scientists. Some of the examples provided by Allchin include the role of gender and race in rationalisation of scientific knowledge. Third, science itself can generate values that can contribute to society, culture and ethics. For example, science as a problem-solving activity is an aspect of science that can have societal uptake that can result in influencing other social systems. Irzik and Nola (2011) consider "simplicity", "empirical adequacy" and "fruitfulness" as examples of epistemic values of science:

> The aims in question are not moral but cognitive. Of course, there are many other aims in science such as consistency, simplicity, fruitfulness and broad scope (Kuhn, 1977); high confirmation, as emphasized by logical empiricists (Hempel, 1965, Part I); falsifiability and truth or at least verisimilitude (i.e. closeness to truth) (Popper 1963, 1975); empirical adequacy (van Fraassen 1980), viability (von Glasersfeld, 1989), ontological heterogeneity and complexity, as emphasized by empiricist feminists like Longino (1997). (p. 597)

Let's take "simplicity" as an epistemic value in chemistry. Hoffman, Minkin and Carpenter (1997) discuss the application of this value in the context of reaction mechanisms in chemistry. In arguing for the role of philosophy in chemistry, they take on what's typically referred to as the "Ockham's razor" generally taken to mean that one should not complicate explanations when simple ones will suffice. The authors state that "the context in which Ockham's Razor is used in science is either that of argumentation (trying to distinguish between the quality of hypotheses) or of rhetoric (deprecating the argument of someone else)" (p. 3–4). Hoffman and colleagues give the example that a definition of reaction mechanism can involve an obvious analogy with the mechanical description of particles. They refer to the work of Rice and Teller (1938) who proposed the "principle of least motion" (PLM) according to which "those elementary reactions will be favored that involve the least change in atomic position and electronic configuration" (p. 489).

In reviewing epistemic aims and values of science for consideration in science education, Erduran and Dagher (2014) drew from philosophy of science accounts to make the case that the epistemic, cognitive and social aims and values of science can be infused into science education. Their review illustrates that it is often rather difficult to disentangle epistemic aims and values from others that have cognitive and social dimensions. Irzik and Nola (2011) argued "aims" and "values" are at times difficult to distinguish because they might serve similar functions in different aspects of science. In order to capture this difficulty in separation of aims and values of science, Erduran and Dagher (2014) proposed a visual representation in the form of a triangle where the epistemic, cognitive and social aims and values are interactive and interrelated. The difficulty in disentangling particular aims and values in chemistry is reflected in the value of objectivity and accuracy in chemistry. David Baird reviews different aspects of objectivity in relation to chemistry. He illustrates how at times, "...objective methods are tied up with economic ends; they involve the combination of productivity and accuracy" (Baird, 2000, p.100):

> This is remarkable because separating the idea of accuracy from that of objectivity runs counter to the intuitive, perhaps even analytical, alignment of accuracy and objectivity. These changes tie instrumental objectivity more closely to productivity and loosen its connection with accuracy. (Baird, 2000, p. 100)

The sense of intermixing of some traditional epistemic aims and values of science with other values such as productivity in the context of chemistry is further exemplified in the work of Bensaude-Vincent who points out that

> The notion of technoscience denotes not only the actual hybridization of science, technology, commercial and industrial interests but also a new ideal of scientific practice where epistemic values (such as truth, simplicity etc.) are explicitly challenged by non-epistemic values such as social robustness, social and economic relevance and sensibility to environment. (Bensaude-Vincent, 2013, p. 336)

In this sense, as an industrial science, chemistry's epistemic aims and values can thus be conflated with other values such as aesthetic values. Aesthetic values can play a role in attracting people to chemical laboratory work, developing new materials, selecting and designing synthetic targets on a theoretical level, interpreting molecular representations, performing chemical experiments, and developing mathematical models (Schummer, 2014). Apart from reliable models for explanation and prediction, chemists are also engaged in the classification of substances and reactions, be it synthetic or analytical. Such pluralism of aims entails a pluralism of values, including not only epistemic values, such as predictive and explanatory values, but also instrumental or technological values, such as the usefulness, practicability and performance of tools and methods. Schummer points to the role of quest for symmetry as an epistemic value in chemistry:

> Mathematical symmetry plays a fundamental role in chemistry to describe crystal structures and molecules, to identify forms of molecular isomerism, to develop quantum-chemical models, to analyze spectroscopic results, and so on. There are even quantum-chemical rules, the Woodward-Hoffman rules, for which Roald Hoffmann received the 1981 Nobel Prize in Chemistry, that predict the products of certain reactions from the symmetry of molecular orbitals. Apart from such routine uses, however, symmetry is also a guiding principle of research by suggesting certain explanations about the natural order of substances or certain synthetic strategies for the design of new products. In these contexts, symmetry functions as an aesthetic principle that can guide or misguide research from an epistemic point of view. (Schummer, 2014, p. 320)

Chemists strive to produce the ideal, aesthetically preferred form of substances. The most prominent and important one is the ideal crystal, which requires tremendous efforts at purification and recrystallisation, without being ever achieved in practice because of remaining impurities and entropy effects (Schummer, 2014). The ideal crystal has perfect translational symmetry such that a small unit represents the whole crystal, which allows for theoretical representation. In a broader sense, as Bensaude-Vincent (2013) argues:

> ...a distinctive feature of chemistry is that making molecules is presented as a means towards an end: it can be healing or killing, bringing health or death, or creating a beautiful architecture. In any case, chemistry is loaded with values such as public good, utility, beauty. (p. 336)

It is beyond the scope of this chapter to present an exhaustive review of the epistemic aims and values of chemistry. The few selected examples address the question of what domain-general and domain-specific aims and values can be identified in chemistry as a domain of science such that school chemistry can be enriched in application of these aims and values. Although some of the aims and values highlighted in this section are relevant for school chemistry, it is vital to consider their curriculum relevance, the cognitive demands placed on the students as well as the teachers' skills in teaching them. A prerequisite to secondary students' understanding of fairly sophisticated concepts such as objectivity and empirical adequacy is the basic awareness that there are epistemic aims and values in science in the first place. Subsequently domain-specific and nuanced examples can be built into learning progressions. A similar argument can be made for pre-service and in-service teachers' learning as well. Most chemistry teachers will be fairly unfamiliar with epistemic aspects of chemistry. Hence, their learning will need to prioritise their basic awareness of the epistemic nature of science, subsequently building on their understanding through more specific chemistry examples.

## 2.3  Practices in Chemistry

Scientists gather and analyse data in order to formulate scientific knowledge. Knowledge discovery and creation in science follows systematic exploration, observation, description, analysis and testing of phenomena and facts within the communication framework of a particular research community with its accepted methodology and a set of techniques (Kwasnik, 1999). Scientific practices such as experimentation, observation and classification all contribute to how scientists generate data. In all branches of science, data underscore the explanations that scientists construct. Irzik and Nola (2011) distinguish between observational and experimental data. While the authors did not unpack this issue in a particular paper, Gurol Irzik provided the following distinction which was reported in Erduran and Dagher's (2014) book:

> *Each observational and experimental data can be expressed in terms of a statement of the form: such and such object has such and such property. For that reason, statements that express observational and experimental data are singular statements, which are different from scientific laws (such as PV = constant) that are expressed in terms of universal statements. Given all this, both observational and experimental data, provided they contain no errors, typically function as evidence for or against theories or hypotheses, they are used in scientific explanations and often called initial conditions and thus constitute part of the corpus of scientific knowledge. In his Logic of Scientific Discovery, Popper discusses these points at some length* (from Erduran & Dagher, 2014, p. 73)

Irzik further explains that observational reports concern the data that are obtained through observation, for example, the data obtained with a telescope on a planet at different times and different locations. Experimental data, on the other hand, are obtained through experiment. For example, the measurements that enabled Boyle to find out about the law that carries his name (as expressed by PV = constant). Broadly speaking, both forms of data could count as observations.

Chemists use data to formulate models that can help explain phenomena (e.g. Suckling, Suckling, & Suckling, 1978; Tomasi, 1988; Trindle, 1984). Philosophers of science often situate models as intermediaries between the abstractions of theory and the concrete actions of experiment (Downes, 1993; Redhead, 1980). They examine explanatory power of models (Cartwright, 1983; Woody, 1995) and the relation of models to theories (Giere, 1991). History of chemistry is replete with various accounts of models. For example, Justi and Gilbert (2000) refer to various models such as the "anthropomorphic" model which a chemical change in terms of the readiness of the components to interact with each other. The "affinity corpuscular" model emphasised the chemical change in terms of atomic affinities. "First quantitative" model introduced the notion of proportionality of reactants for chemical change to occur. The "mechanism" model began to outline steps in a chemical reaction. The "thermodynamics" model drew attention to the role of molecular collision (with sufficient energy) in chemical change. The "kinetic" model introduced the idea of frequency of collisions of molecules. The "statistical mechanics" model relied on quantum mechanics and identified a chemical reaction as motion of a point in phase space. The "transition state" model provided a link between the kinetic and thermodynamic models by merging concepts of concentration and rate.

Along with modelling, classification and experimentation play important roles in chemistry. A good example from chemistry of a classification system is the periodic table of elements (Hjorland, Scerri & Dupre, 2011). When the periodic table was first proposed, there was already a body of knowledge about individual elements based on knowledge such as atomic weight (Scerri, 2006). It was observed that elements could be arranged in a systematic order according to atomic weight and this would show a periodic change of properties. This early periodic table proved to be a very useful tool, leading to the discovery of new elements and a new understanding of already-known elements. Experimentation, on the other hand, has a long history in philosophy of science (e.g. Latour & Woolgar, 1979; Shapin & Schaffer, 1985). Radder (2009) outlines two primary features of experimentation: intervention and reproducibility. In order to perform experiments, experimenters have to intervene actively in the material world; moreover, in doing so they produce all kinds of new objects, substances, phenomena and processes. Radder explains that experimentation involves the material realisation of the experimental system as well as an active intervention in the environment of this system. Hence, a central issue for a philosophy of experiment is the question of the nature of experimental intervention and production and their philosophical implications.

The epistemic practices such as observation, classification and experimentation are mediated by cognitive and social practices such as reasoning, argumentation and social certification of ideas. Erduran and Dagher (2014) produced a visual tool referred to as the "Benzene Ring Heuristic" (BRH) (see Fig. 2.2). BRH uses the analogy of the benzene ring to summarise scientific practices. Each carbon atom around the ring and the diffuse pi bonds represents the social contexts and practices that apply to all of these aspects. The cognitive, epistemic and social aspects of science are interrelated and influence one another. The ring structure represents the "cloud" of cognitive and social practices that mediate the epistemic components such as models and explanations.

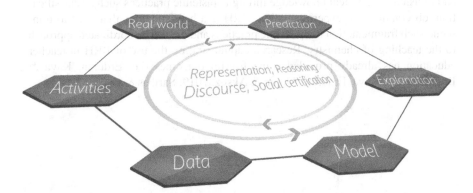

**Fig. 2.2** Benzene Ring Heuristic of scientific practices. (From Erduran & Dagher, 2014, p. 82)

BRH articulates how scientists use data originating from the real world to generate models, explanations and predictions. In a sense, the heuristic highlights the mechanisms for how the knowledge growth occurs. In school chemistry, the activities of experimentation, classification and observation tend to be covered in a disconnected fashion that doesn't necessarily lead to modelling practices. Many textbooks do not bring forth the epistemic practices of science from the readers' point of view. For instance, in an analysis of chemistry textbooks from Lebanon, BouJaoude, Dagher and Refai (2017) identified some missed opportunities where the epistemic features of chemistry could have been covered:

> ..the chemistry textbook provides sufficient detail in terms of conceptual knowledge as determined by the Lebanese curriculum objectives but misses the opportunity to introduce to students to the epistemological and social aspects of science even though such opportunities could be easily seized upon. (p. 91)

Furthermore, although many chemistry textbooks might include specific components of scientific practices such as reference to the periodic table as a classification system, many textbooks do not present scientific practices in a coherent and coordinated way. In a recent large-scale Australian study, for instance, McDonald (2016) indicated that the top three textbooks currently being used by Australian schools were fairly limited in how they represented nature of science. In subsequent work, McDonald (2017) states:

> Opportunities to consider the NOS categories of scientific practices and methods and methodological rules were primarily provided whilst examining science inquiry activities. Unfortunately scientific practices were generally portrayed as isolated activities, and many investigations included in the chapters followed a recipe-style format. Although the methods included did not reinforce the myth of the scientific method and a variety of methods were often presented to students, no explicit consideration of how these methods contribute to knowledge in the field was considered. (McDonald, 2017, p. 112)

In school chemistry, recipe-following, a significant problem often referred to as the "cookbook problem" (van Keulen, 1995), is often disguised as chemical experimentation. Chemistry, the science of matter, is not driven by recipes nor by data collection and interpretation alone. Chemists contribute to the development, evaluation and revision of chemical knowledge through epistemic practices such as classification, observation and experimentation. BRH is a heuristic aims that aims to transcend such fragmentation of epistemic practices and present a coordinated approach to the teaching of chemistry. Research capitalising on the use of BRH in teacher education has already provided some fruitful outcomes (i.e. Erduran, Kaya & Dagher, 2018; Kaya, Erduran, Akgun & Aksoz, 2019; Saribas & Ceyhan, 2015).

## 2.4   Methods in Chemistry

School science typically projects a definition of the scientific method as a linear process that progresses from formulation of a hypothesis to the conduct of an experiment to reach the findings. As such, it is often presented as an unproblematic and stepwise process. One of the shortcomings of promoting "the scientific method" in such a popular form is communicating that there is indeed a "uniform, interdisciplinary method for the practice of good science" (Cleland, 2001, p. 987). The emphasis on "the scientific method" also contributes to the perception that doing credible scientific work necessitates using this method. This leads to the false conclusion that scientists who do not use experimental methods are not likely to arrive at trustworthy knowledge. For example, if chemists are merely observing colour changes in a reaction, such observations may not be deemed "scientific enough" although macroscopic properties are an essential component of chemistry and they complement microscopic properties. Sometimes the identification of colour can be decisive in making inferences about a reaction (e.g. precipitation reactions). This is a case of non-manipulative observation that can help address a hypothesis about a reaction type. It is about observation of the parameter of colour and how inferences on colour changes can point to a reaction type.

Brandon (1994) discussed different methodologies in science. Even though Brandon explores the idea of experiment in the context of biology, his work is relevant for other science domains such as chemistry. Brandon (1994) depicts two ways in which experiments are usually contrasted: contrast with observations and contrast with descriptive work. Critical to the contrast between experiment and observation is the occurrence of manipulation that he defines in a restricted sense. This restricted sense of manipulation in this case rules out interventions that do not alter the phenomena. He gives the example of dissection as a non-example of manipulation as it involves the making visible of otherwise invisible phenomena. Thus manipulation in the context of this discussion "involves the deliberate alteration of phenomena" (Brandon, 1994, p. 61). An example of manipulation in the restricted sense would be an instance where independent variables are changed to allow the documentation of their effect on dependent variables.

In terms of the contrast of experiment with descriptive work, a key factor to the contrast is whether a hypothesis is being tested or whether the values of parameters are being measured. Parameter measures may demand considerable manipulation but may or may not involve the testing of hypotheses. Brandon (1994) gives the following example. If biologists are interested in finding out whether a given herbivore can exert a selective factor for a population of plants, the herbivore (serving as independent variable) and variables pertaining to its effects on plant survival and reproduction would be introduced to an experimental plot, without necessarily posing a hypothesis. Brandon's examples illustrate that not all experiments involve hypothesis testing and that not all descriptive work is non-manipulative. He represents the connections between experiments and observations in terms of a two-by-two table reproduced here. The nature of the investigation (experiment/observation)

**Table 2.1** Experiment/observation, manipulation/non-manipulation and descriptive/experimental categories

| Experiment/ observation | Manipulate | Not manipulate |
|---|---|---|
| Test hypothesis | Manipulative hypothesis test | Non-manipulative hypothesis test |
| Measure parameter | Manipulative description or measure | Non-manipulative description or measure |

Reproduced from Brandon, 1994, p. 63

**Table 2.2** Experiment/observation, manipulation/non-manipulation and descriptive/experimental categories with examples related to the periodicity of elements

| | Manipulate | Not manipulate |
|---|---|---|
| Test hypothesis | Manipulative hypothesis test | Non-manipulative hypothesis test |
| | For example, *Crookes' study of gases* | For example, *De Boisbaudran's discovery of gallium* |
| Measure parameter | Manipulative description or measure | Non-manipulative description or measure |
| | For example, *Rutherford's artificial transmutation of elements* | For example, *Mendeleev's prediction of gallium* |

From Erduran and Dagher (2014, p. 102)

is related to whether or not (a) it involves manipulation and (b) hypothesis testing or parameter measure (see Table 2.1). According to his analysis, one can think in terms of experiment and non-experiments/observations relative to descriptive versus experimental methods. Erduran, Cullinane and Wooding (2019) used Brandon's matrix as a framework to investigate national chemistry examination papers from England. The examination items from two examination papers of a leading examination board were classified according to these categories, and patterns on the marking were traced. The results indicated that for both papers, non-manipulative parameter measurement was the method assessed at a higher percentage. In both papers, manipulative hypothesis testing was the category with the lowest percentage of items or questions. Furthermore, the mark allocation was the highest in both papers in the non-manipulative parameter measurement category.

Erduran and Dagher (2014) referred to the philosophy of chemistry literature in highlighting the diversity of methods in science (see Table 2.2). They drew on the work of Scerri who describes how Mendeleev predicted the existence of the element gallium (or eka-aluminium) through a non-manipulative description coupled with quantitative reasoning about atomic weights:

> *Mendeleev could interpolate many of the properties of his predicted elements by considering the properties of the elements on each side of the missing element and hypothesizing that the properties of the middle element would be intermediate between its two neighbors. Sometimes he took the average of all flanking elements, one on each side and those above and below the predicted element. This interpolation in two directions was the method he used to calculate the atomic weights of the elements occupying gaps in his table, at least in principle.* (Scerri, 2007, p. 132)

Scerri states that it was the French chemist Emile Lecoq De Boisbaudran who sub-sequently "worked independently by empirical means, in ignorance of Mendeleev's prediction, and proceeded to characterize the new element spectroscopically" (Scerri, 2007, p. 135). De Boisbaudran was testing the hypothesis of the existence of a new element by spectral analysis of an ore and managed to isolate gallium through this method. The manipulative aspect of some chemical methods includes (a) Crookes' study of gases where pressure and voltage were used as variables in spectroscopic study of elements (e.g. Scerri, p. 251) as an example of manipulative hypothesis testing and (b) Rutherford's artificial transmutation of elements through bombardment of nuclei with protons (e.g. Scerri, p. 253) as an example of manipu-lative description. All together these methods, along with numerous others, contrib-uted to the collective and eventual depiction of elements.

Identifying features of investigations that distinguish them as belonging to one of the four quadrants does not necessarily mean that all investigations in any one quad-rant are carried out in the same way or follow an algorithm. All it means is that these methods share some distinctive features (e.g. involve hypothesis testing or conduct-ing observations). Erduran and Dagher (2014) noted that it is important then to point out that they are not advocating the replacement of "the scientific method" with "four scientific methods". Rather, the point of referring to the matrix is to emphasise the range of ways in which investigations can be set up to address different research questions. Furthermore, the use of heuristics such as the two-by-two tables can pro-vide meta-tools for communicating how different science domains or indeed par-ticular examples within a domain of science might be employing methods. Erduran and Dagher (2014) visually represented the dynamics of interaction that needs to take place in how scientists coordinate evidence from different methodological approaches (see Fig. 2.3). In other words, problems in science are very unlikely to be solved through just one method alone. Rather, several methods are used to gener-ate data that need to be then coordinated to generate scientific theories and models.

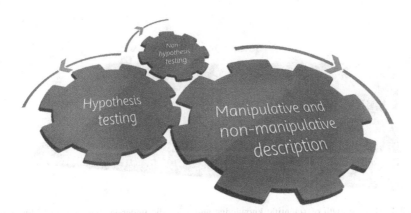

**Fig. 2.3** Diversity of methods working synergistically. (From Erduran & Dagher, 2014, p. 101)

## 2.5  Knowledge in Chemistry

Science has had a tradition of producing knowledge that helps explain and predict physical and natural phenomena. While some knowledge is in the form of theories, such as the atomic theory, others might be best characterised as models and laws, for instance, molecular models and the law of conservation of mass. There is vast amount of research literature on each of these knowledge forms (e.g. Cartwright, 1983; Christie & Christie, 2003; Giere, 1991; Machamer & Woody, 1992; Suckling et al., 1978). Erduran and Dagher (2014) argued that theories, laws and models (TLM) are different forms of scientific knowledge that work together to produce scientific knowledge. Erduran and Dagher (2014) represented TLM and growth in TLM visually (see Fig. 2.4). While the visual heuristic presented in Fig. 2.4 emphasises the components of scientific knowledge and its growth, Fig. 2.2 which was referred to previously illustrates some of the mechanisms and interactions that mediate how the various forms of scientific knowledge need to be coordinated in a finer level of detail. The instructional adaptations of BRH can be used to get the students to think about, for instance, how a model of the atom is related to the "real world" or how the atomic model can help predict chemical behaviour. As Erduran and Dagher (2014) argued, the visual tools are generative in helping guide the discussion of various aspects of scientific knowledge and practices.

Erduran (2014) has proposed the example of the atom to illustrate how TLM works. TLM illustrates how different knowledge forms contribute to knowledge and how knowledge growth occurs in science. It also can illustrate how scientific knowledge can vary in terms of the nature of theories, laws and models in different domains of science. For example, structure of matter is explained with different types of knowledge that are the atomic theory, the periodic law of elements and molecular models. While knowledge in physics would also have theories, laws and models, the precise nature of them may be specific. For example, "law" might have a specific meaning in chemistry as compared to physics (e.g. Christie & Christie, 2003). If a particular TLM at a certain point in time cannot explain a phenomenon,

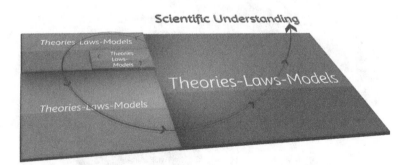

**Fig. 2.4** TLM, growth of scientific knowledge and scientific understanding. (From Erduran & Dagher, 2014, p. 115)

a paradigm shift might occur. In chemistry, the shift to Lavoisier's theory of oxygen from the phlogiston theory is an example of a paradigm shift (Erduran, 2014).

If we continue to unpack the atom example, each stage in the formulation of chemical understanding around the atom can be represented by a separate plane, for example, the ancient Greek depiction of four elements (earth, water, fire, air), and Lavoisier's conception of elements would constitute separate "planes" that have represented knowledge accumulation over time. When TLM couldn't account for observed evidence and the entire paradigm had to change, scientists began to formulate a new plane of knowledge. Kuhn himself used the example of atomic theory to illustrate the concept of incommensurability (Kuhn, 1962/1970, p.85) meaning that different paradigms would not map onto each other due to their fundamentally different paradigmatic assumptions. The change from properties of matter such as colour to quantitative analysis of reactions could not possibly be mapped to the same knowledge plane. The broader theoretical framework, the regularities and patterns in the behaviour of matter and the representations of properties or reactions worked in unison to provide an overall paradigm. Altogether, TLM brings coherence to the various forms of scientific knowledge, illustrates how they are related and also accounts for how TLM grows as evidence accumulates. TLM grows in time as evidence accumulates leading to scientific understanding at every stage. However, there may be points in time when the TLM are no longer able to account for the evidence observed and hence a new set of knowledge forms are needed, leading to a new paradigm, referred to by Thomas Kuhn as a "paradigm shift". A new paradigm could be represented as a new "plane" where the knowledge accumulation process begins and continues.

Theories, laws and models work together in the growth of scientific knowledge. Often, school science presents different forms of scientific knowledge without articulating them coherently. With the atom example, Erduran (2014) states that this means that "even though the atomic theory, the atomic model and periodicity as a pattern in elements are introduced to students, there is hardly ever any indication of what these forms of knowledge are about or how they interact in leading to understanding the structure and function of matter" (p. 40). She further raises questions such as "what is the relationship between the atomic theory and the atomic model? Are there law-like regularities that contribute to our understanding of the atomic model and how are they related to what makes a model?" (p. 40). In order to understand the nature of scientific knowledge, it is important to differentiate between the different forms of scientific knowledge. While theories, laws and models differ from each other, particular definitions and varieties of each knowledge form are equally complex. For example, when we consider theories, we see that there are different types of theories such as centre, frontier and fringe levels of theories (Dutch, 1982). Likewise, there are different descriptions of models (e.g. Bruner, 1966; Giere, 1991). For instance, while Bruner (1966) argued enactive, iconic and symbolic or conceptual models, Giere (1991) proposed scale, analog and theoretical models. Similarly, laws as the other type of knowledge have different versions (Christie & Christie, 2003).

Understanding the nature of scientific knowledge can be challenging to students and teachers alike. For example, students tend to think that laws are the proof of theories (McComas, 1998). Although scientific knowledge including theories, laws and models is covered in science classes, there is rarely an emphasis on explaining the relationships among different types of scientific knowledge and their contributions to the growth of scientific knowledge. In their analysis of Lebanese textbooks, BouJaoude et al. (2017) observe that

> *Even though modeling was discussed, it was not introduced as a fundamental scientific practice and models were not discussed as significant forms of scientific knowledge. Yet again, discussing modeling and the nature of scientific models was missed.* (p. 91)

A further shortcoming of school chemistry is that the curriculum does not represent key concepts such as the atomic theory to students in relation to the associated models and laws or how they possess explanatory and predictive power that leads to chemical understanding. Rather, students are typically introduced to disparate pieces of knowledge (e.g. "the atom consists of a nucleus and electrons"). Whenever knowledge from different paradigms are presented (e.g. Dalton's vs Bohr's atom), there is rarely any lead into the historical development of the associated ideas. How, then, could heuristics such as the one represented in Fig. 2.4 help teachers and students in understanding chemistry knowledge? Duschl and Erduran (1996) had proposed some pedagogical tools (e.g. writing heuristics and visual representations) for teachers to focus on highlighting the structure and the growth of scientific knowledge. Such pedagogical tools might help develop students' and teachers' metacognitive understanding of the nature of scientific knowledge. In a similar vein, the heuristic in Fig. 2.4 is a "meta-tool" that highlights the significance of understanding what constitutes chemical knowledge. Visual heuristics potentially can give teachers and students a sense of the progression of ideas, how ideas change over time and how ideas can at times be abandoned altogether and replaced by new ones. In short, such tools can help understanding of different knowledge forms in chemistry and how they work together to produce understanding in chemistry over time (Erduran, 2014).

## 2.6  Applying the Epistemic Core to Chemistry Concepts

The four categories of scientific aims and values, practices, methods and knowledge from Erduran and Dagher's (2014) framework constitute the core epistemic idea sets related to the epistemic features of science in general. These categories can be exemplified in the context of a particular domain such as chemistry. They represent a comprehensive and inclusive set of epistemic ideas that underpin how knowledge production and development in science occur. In the rest of this book, we will refer to these categories collectively as the "epistemic core". The terminology allows us to have a quick reference to the key epistemic aspects of chemistry that we may target in chemistry education. The epistemic core provides a simple set of four

**Table 2.3** Example concepts for illustrating the epistemic core in the context of acids and bases

| Aims and values | Methods |
|---|---|
| Empirical adequacy | Non-manipulative description |
| Critical examination | |
| **Practice** | **Knowledge** |
| Modelling | Models |
| Classification | |
| Argumentation | |

categories that can be populated with various examples. It is comprehensive and inclusive in terms of capturing the various epistemic dimensions that can be considered in relation to a science. In considering the epistemic core in relation to an educational example, we focus on some selected concepts about each aspect (i.e. aims and values, methods, practices and knowledge) as indicated in Table 2.3. Suppose we focus on empirical adequacy and critical examination for aims and values; non-manipulative description for methods; modelling, classification and argumentation for practices; and models for chemistry knowledge. Clearly, not all possible aspects of each category can be included all at once. Some decision-making will be called for in making choices around the particular instances. Such decision-making will need to be consistent with curriculum goals and the cognitive ability of the students or teachers in question. If various combinations get covered across time, it is anticipated that teachers' and students' understanding of the epistemic core will be enriched.

A typical activity in acid-base chemistry in lower secondary school is the classification and observation of some properties of everyday acids and bases. Students are asked to experience everyday materials such as lemon juice, vinegar and soap and consider their properties as being properties of acids and bases such as "sour" for acids and "slippery" for bases. This task involves classification of substances on the basis of properties. What is typically not pursued is asking the students to go further and try to explain why acids and bases have such properties. In previous work, Erduran (1999) asked 12-year-old students to draw pictures of what makes an acid have its properties and likewise for the base. Here the intention is to encourage the students to generate their own models of acids and bases from the sensory data that they obtain. The representations that are produced include rather spiky or jagged figures for acids and soft, rounded shapes for bases (see Fig. 2.5). When they are asked to predict what would happen if you mixed acids and bases together, they try to combine the features of the shapes where the effects of either are cancelled out. For instance, the spikes of the acid penetrate the holes of the base. There is thus prediction of a model of neutralisation.

If students are asked not only to classify and observe but also to model the components of acids and bases to explain the observed properties, such an activity is enriched with an epistemic purpose. Now students can discuss and debate how the different representations capture or not the observed properties. They can also make

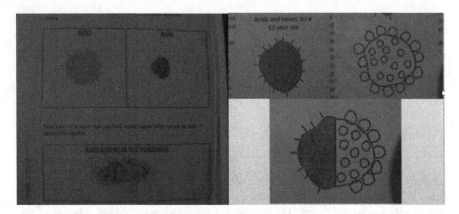

**Fig. 2.5** A 12-year-old student's models of acids and bases (Erduran, 1999)

predictions about what happens if you were to put the two substances together. In terms of the use of the Benzene Ring Heuristic (see Fig. 2.2), there would be plenty of opportunities to contextualise acids and bases relative to real world (i.e. everyday substances), activities (i.e. observation, classification), data (i.e. sensory experiences of taste and touch), modelling (i.e. pictures to explain properties), explanation (i.e. why acids are sour and bases are slippery) and prediction (i.e. an account of neutralisation). The students are asked to classify for a purpose which is to explain observed properties. The methodological approach involves non-manipulative description. The students are simply describing properties and trying to account for them without manipulating any variables.

It can be argued that the student's representations of acids and bases from Fig. 2.5 have historical authenticity. It was common practice in seventeenth-century chemistry to describe acids and bases on the basis of shape. Nicolas Lemery, a French chemist

> *explained chemical and physical properties by shape. Acids had sharp spikes on their atoms, accounting for the pricking sensation that exert on the skin. Alkalis were highly porous bodies into which the spikes of the acids penetrated and were broken or blunted in producing neutral salts.* (Leicester, 1971)

Thus, the example could be suggestive of students' engagement in authentic chemical inquiry where ways of thinking and reasoning in chemistry are modelled and students generate chemical knowledge themselves. In the social environment of the classroom, students will produce different models. There can be debates around which models explain the observed sensory data and which do not. As such, students can be engaged in argumentation activities where they produce evidence and justifications for the claims that they are making in constructing their models. Such discussions can be mediated by the teacher, and they would involve critical examination of coherence between observations and claims underpinning the models, e.g. the spikes have sharp-pointed edges that account for the pricking sensation. In this sense, there's empirical adequacy of the constructed models.

The preceding example illustrates what is possible to target in terms of teaching some example concepts related to the epistemic core of chemistry. If we focus on one category of the epistemic core, further themes can be unpacked. For example, if we take the "knowledge" category in the context of acid-base chemistry, there's a great deal of potential for covering themes such as development of chemical knowledge including model evaluation and revision as described in Sect. 2.5. The definition of acids and bases on the basis of shape as advanced by Nicolas Lemery leads to subsequent proposals on how to define acids and bases. A brief and crude account of the history of acid-base chemistry illustrates how this issue can be picked up across different stages of schooling where other models are introduced. From a curriculum perspective, the issue becomes one of how students' existing models are transformed to coincide with more contemporary accounts of acids and bases. For example, towards the end of the nineteenth century, Svante Arrhenius classified a compound as acid or base according to its behaviour when it is dissolved in water to form an aqueous solution. He suggested that a compound be classified as an "acid" if it contains hydrogen and releases hydrogen ions, $H^+$ (see Table 2.4). Likewise, a "base" was defined as a compound that releases hydronium ions, $OH^-$, in a solution.

Subsequently, Johannes Brønsted and Thomas Lowry proposed a broader definition of acids and bases, which assigned an intrinsic property to acids and bases, independent of their behaviour in water. The new model was formulated based on observations that substances could behave as acids or bases even when they were not in aqueous solution, as the Arrhenius model required. In the Brønsted-Lowry model, an acid is a hydrogen donor and a base is a hydrogen acceptor. There is no requirement for the presence of water. Acid-base chemistry took yet another turn when the centrality of hydrogen in both the Arrhenius and Brønsted-Lowry models was challenged. Gilbert Lewis formulated another definition of acids and bases. He emphasised the centrality of electrons in characterising acid-base behaviour. Lewis considered that the crucial attribute of an acid is that it can accept a pair of electrons and of a base that it can donate a pair of electrons. In the context of the Lewis model, electron donation results in the formation of a covalent bond between the acid and the base.

| Table 2.4 Models of acids and bases | Models | Chemical equation |
|---|---|---|
| | Arrhenius | $HA(aq) \rightarrow H^+aq + A^-aq$ <br> Acid |
| | | $BOH(aq) \rightarrow B^+(aq) + OH^-(aq)$ <br> Base |
| | Brønsted-Lowry | $HA(aq) + B \rightarrow HB^+(aq) + A^-(aq)$ <br> Acid      Base |
| | Lewis | $2H^+ \quad + :O^{2-} \rightarrow \ 2O{-\!}H^-$ <br> Acid      Base |

**Table 2.5** Chemistry examples of theories-laws-models (TLM)

| Form of knowledge | Atom | Gases | Bonding | |
|---|---|---|---|---|
| Theory | Atomic theory | Kinetic energy theory | Molecular orbital theory | Lewis theory of bonding |
| Law | Periodic law | Boyle-Mariotte law | Octet rule | Coulomb's law |
| Model | Atomic model | $PV = nRT$ | VSEPR model | Lewis model |
| TLM explain | Structure of matter | Behaviour of gases | Geometry of molecules | Ionic bond |

The use of the different models of acids and bases creates an opportunity to illustrate not only the nature of models but also the criteria that drove model evaluation. For example, while in the Arrhenius definition acidity was defined relative to behaviour when an acid is dissolved in water to form an aqueous solution, it was considered independently of an acid's behaviour in water in the Brønsted-Lowry model. The empirical adequacy of each model can be discussed along with the constraints of different models in explaining properties. There is vast research on problems associated with the teaching and learning of chemistry although they are promoted in the chemistry curriculum. For example, many students do not understand models as tentative representations but rather consider them as copies of reality (Grosslight, Under, Jay, & Smith 1991). Textbooks often do not make clear distinctions between chemical models and present inaccurate "hybrid models" (Justi & Gilbert, 2000). The consideration of the epistemic core in acid and base chemistry can be extended to other topics. For example, the atom, gases and bonding can be the context for organising relevant aspects of the epistemic core. In one scenario, theories or laws can be brought to the foreground, and in another the interconnection of the theories, laws and models can be discussed (see Table 2.5). Decisions on what topic to choose and how to organise it for what purpose will depend on the goals of the teacher relative to the curriculum expectations for the particular age group of students. Similarly, teacher educators can adapt the use of various examples of TLM and BRH depending on what instructional approaches they set for pre-service and in-service teachers' learning.

In the example of ideal gas laws, the relationships between different variables such as pressure, volume and temperature can be examined (i.e. experimentation in terms of scientific practices; manipulative hypothesis testing in terms of methods). This topic lends itself to illustrate different methods in chemistry. Furthermore, it can be linked to practices of chemistry because explanations and predictions of the behaviour of gases are called for. Representations such as graphs based on data can be drawn, and alternative representations and explanations underpinning these representations can be discussed. BRH can help structure and organise the concepts involved. Repeated coverage of topics at the same grade level structured with an epistemic "frame" is likely to instil in students ways of thinking that are implicit in the design of the visual tools such as the BRH.

Similar epistemic framing of topics can be considered across grade levels as well. For example, in the case of acid-base chemistry, an activity that could be used at more advanced levels could be the following. Students are provided with equations of Arrhenius, Brønsted-Lowry and Lewis models of acids and bases. Then they are given a set of acids and bases and asked to group them and derive their particular equations. Here there would be classification and modelling in order to explain and predict their behaviour. Layering these activities with some epistemic structure and purpose would delineate this activity as different from what is typically done in chemistry lessons. For example, the participants can be probed through questions such as "how did your classification help you think about the model of acids and bases?" and "how does your model explain and predict the chemical reaction?". The sense of intermixing epistemic aims and values such as objectivity and simplicity is coupled with other non-epistemic values such as economic relevance (Bensaude-Vincent, 2013) and aesthetics (Schummer, 2014). Apart from reliable models for explanation and prediction, chemists are also engaged in the classification of substances and reactions, be it synthetic or analytical. Such pluralism of aims entails a pluralism of values, including not only epistemic values, such as predictive and explanatory values, but also instrumental or technological values, such as the usefulness, practicability and performance of tools and methods. In summary, depending on the curriculum objectives and the instructional goals, any aspect of the epistemic core can be populated with different chemistry topics, and their increasing complexity can be captured in careful sequencing in teaching or teacher education to meet the demands of the students or teachers, respectively.

## 2.7 Implications for Chemistry Education

Shifting the emphasis in chemistry education from learning concepts such as acids and bases to learning other aspects of chemistry such its epistemic core requires recalibration of what are expected outcomes in terms of students' and teachers' learning. Table 2.6 illustrates some examples of student learning outcomes that can be expected of secondary students. The details of each learning outcome can be specified in an appropriate manner for the age group. For example, as illustrated with the acids and bases scenario, it may be appropriate for students' models at age 12 to be based on representations of shape of ingredients. Progressively once the students cover other accounts of acids and bases, they would then be in a position to evaluate the criteria for model revision. Hence while the construction of models on the basis of properties of substances might be a relevant learning goal for lower secondary, for upper secondary level, there may be emphasis on model evaluation and revision on the basis of reactivity with metals, for example.

In terms of teachers' learning, a similar approach can lead to specification of what learning outcomes can be targeted for teachers from different career stages. For example, a goal for pre-service teacher education could be that pre-service teachers understand that chemistry has different epistemic aims and values, with

**Table 2.6** The epistemic core and potential student learning outcomes

| Aspect of epistemic core | Potential student learning outcomes |
| --- | --- |
| Aims and values | Students will understand that chemistry is guided by certain epistemic aims and values such as objectivity, accuracy and empirical adequacy |
| | Students will appreciate that sometimes chemistry's epistemic aims and values may be conflated with aesthetic values and economic demands |
| Practices | Students will be able to classify substances and construct models to explain their properties |
| | Students will engage in debates about how and why chemical models point to particular explanations and predictions |
| Methods | Students will recognise a range of methods that chemists use some of which require manipulation of variables and hypothesis testing and some are based on non-manipulative descriptions |
| | Students will understand that chemistry relies on the coordination of evidence obtained through different methods |
| Knowledge | Students will differentiate between theories, laws and models in chemistry |
| | Students will understand that knowledge in chemistry develops through evaluation and revision of theories, models and laws which need to have coherence across them |

examples of objectivity and accuracy, while a more experienced teacher might be expected to have a more nuanced understanding of such aims and values being conflated with aesthetic values and economic demands. Furthermore, pre-service teachers would then be able to transform such understanding into teaching practice through the use of different teaching strategies such as group discussions and questioning.

However, the distinction in the expertise between early to mid-career and experienced teachers is not always straightforward. For example, Loughran (2006) reported that when experienced science teachers change schools or have to teach unfamiliar content, they resume the role of the novice science teacher again. Hence, it would be expected that novice or experienced teachers would have similar issues in terms of understanding unfamiliar themes such as epistemic aspects of their subject domain. There is also a further factor related to the difficulties that teachers may face in transforming their learning into teachable content. Chapter 3 explores the implications of the infusion of the epistemic core of chemistry in pre-service teacher education. If the epistemic core is to be integrated effectively in teachers' learning, research evidence on what their background epistemic beliefs are and how teachers learn are important to consider. Hence, the chapter will review empirical evidence on teachers' learning and epistemic beliefs. Subsequently in Chap. 4 the design of an example teacher education project will be described. The project included a dimension of intervention where pre-service teachers were taught a module designed to teach the epistemic core. Hence empirical evidence from pre-service teachers' drawings and verbal statements will be presented in Chaps. 5 and 6 to illustrate the impact of their engagement in the teacher education sessions.

## 2.8 Conclusions

The chapter aimed to provide a framework for defining some core epistemic categories that can help focus the inclusion of examples from chemistry in chemistry education. Erduran and Dagher's (2014) work was used to illustrate the categories of epistemic aims, values, practices, methods and knowledge, and examples from philosophy of chemistry literature were used to show how the epistemic core can be situated in chemistry. The "epistemic core" idea provides focus for organising the content to be drawn from philosophy of chemistry. From a pedagogical point of view, it is simple enough to remember four distinct categories of ideas. Each category of the epistemic core in turn is supplied with a visual tool that can help organise and structure further epistemic themes which can be unpacked with reference to philosophy of chemistry. Examples of the epistemic core were provided with reference to some chemistry topics, and the implications for secondary students' potential learning outcomes were discussed. Overall, the "epistemic core" and the related visual tools provide some structure and educationally relevant focus for how epistemic themes from philosophy of chemistry can be imported and transformed for the purposes of chemistry education.

The epistemic core idea has both domain-general and domain-specific assumptions. In the sense that it is about four categories that can apply to all sciences, it is domain-general. In the sense that it can be populated and instantiated with different disciplinary concepts, it is domain-specific. In some cases the instances can be fairly uniform across different science or topics within a science. For instance, in terms of aims and values, "objectivity" and "empirical adequacy" are fairly universal in all sciences, not only in chemistry. In contrast, some categories such as the knowledge category may invite more domain-specific considerations. For example, models may have some distinct properties in how they are considered in chemistry. As Justi and Gilbert (2000) highlighted, there may be particular references to "anthropomorphic", "affinity corpuscular", "thermodynamics", "kinetic" and "transition state" models that are specific to models in chemistry. History of chemistry can potentially help chemistry educators in identifying the progression of particular aims and values, practices and methods as well as knowledge features of chemistry. Such historical context can potentially enrich the coverage of the epistemic core.

The question remains as to how we, as chemistry teacher educators and researchers, can support school chemistry to ensure that nuanced chemistry examples can be taught and learnt effectively. The specific chemistry examples presented in this chapter are essentially the destination of where we would like to see the teachers and students alike in school chemistry. The overall framework of the epistemic core and the particular examples provide illustrations of the eventual desired outcomes. They do not provide us with any guidelines on how to get there nor what the current state of teachers' and students' knowledge might be about them so that we can build on their ideas. As a significant body of research evidence illustrate, teachers' learning is complex and is mediated by many factors including their epistemic beliefs (see Chap. 3). Furthermore, we do know that chemistry teachers and students have very limited backgrounds in epistemological perspectives of chemistry. Limited exposure in chemistry education to philosophy of chemistry at large is compounded by the fact that chemistry textbooks are also fairly limited in delivering the epistemic core of chemistry to the readers. As McDonald (2017) and BouJaoude et al. (2017) illustrate in different national contexts, although textbooks have related content, they often miss out on the opportunity to present a coherent and coordinated account of the epistemic features.

The aim of this book is not to import philosophy of chemistry into chemistry education for its own sake nor is it to turn chemistry education into a course on philosophy or history of chemistry. Rather, our interest lies in the integration of some epistemic themes to ensure that teachers and students of chemistry are better equipped in making sense of how knowledge and knowing in chemistry work. Chapter 1 outlined some of the potential contributions of the epistemic core to students' metacognition, critical thinking skills and epistemological commitments. On the one hand, we are guided by a pragmatic concern in that as teacher educators, we are dealing with chemistry teachers and students who have very limited historical and philosophical background on chemistry. Given such limited background, and if the destination is to engage secondary chemistry teachers and eventually students in epistemic thinking, then starting with complex chemistry examples from philosophy of chemistry may not be optimal even though their acquisition is the final destination. When teachers and students do not have any awareness of meta-perspectives on science including chemistry, the first step in this process is to develop such awareness at a fairly simple level embedded with some chemistry examples that become increasingly more sophisticated in subsequent iterations as they begin to internalise a meta-perspective on the epistemic aspects of chemistry.

A potential difficulty in making the ideas covered in this chapter a reality concerns teachers' subject knowledge. Teachers might have had sufficient yet incomplete subject knowledge of chemistry which may hinder their use of chemistry examples in unpacking the epistemic core. Furthermore, if they also have limited understanding of the nature of their subject (e.g. meta-level understanding of the aims, values, practices, methods and knowledge), it is questionable if the beginning point of teachers' learning should be a fairly detailed and sophisticated domain-specific approach. Beginning with a baseline of teachers' own ideas about how science works might prove more fruitful in subsequently facilitating their progression towards more nuanced understanding of epistemic categories and their instances in chemistry. However, we should draw a contrast with the implications for students' learning which we are not directly addressing in this book. Hypothetically speaking the epistemic and the conceptual aspects of chemistry may potentially be introduced to students at the same time. However, these issues regarding the sequence of what gets introduced first and progresses in the learning trajectories of teachers and students demand empirical answers. As such, they point to a future research agenda.

The consideration of aims, values, practices, methods and knowledge in unison provides a justified and coherent epistemic framework for chemistry education. The "epistemic core" idea assumes a coordinated approach in terms of the means (e.g. practices and methods), the reasons (i.e. aims and values) and outcomes (i.e. knowledge) of chemical inquiry. Whether the focus is on analysis or synthesis in chemistry, there will be epistemic underpinning of the domain. Without a sense of the epistemic justifications that underscore chemistry, both students and teachers are left in the dark about why chemists do chemistry, how they do it and what they produce as a consequence. Hence understanding the epistemic core of chemistry is fundamental to justifying, from students' and teachers' perspective, why and how chemistry works. Without any awareness of and engagement in the epistemic core, the process of engagement in learning chemistry itself can be said to be based on dogma where students and teachers are presented final versions of chemistry knowledge (e.g. Duschl, 1990) without having any understanding of where such knowledge comes from and why it should be believed in. If chemistry education does not prioritise understanding of the evidence and the justification of the reasons, the processes and the outcomes of knowledge production chemistry education will be far from scientific in its ethos.

# References

Allchin, D. (1999). Values in science: An educational perspective. *Science & Education, 8*(1), 1–12.

Baird, D. (2000). Analytical instrumentation and instrumental objectivity. In N. Bhushan & S. Rosenfeld (Eds.), *Of minds and molecules* (pp. 90–114). Oxford, UK: Oxford University Press.

Bensaude-Vincent, B. (2013). Chemistry as a technoscience? In J. P. Llored (Ed.), *The philosophy of chemistry: Practices, methodologies and concepts* (pp. 330–341). Newcastle upon Tyne, UK: Cambridge Scholars Publishing.

Bhushan, N., & Rosenfeld, S. (Eds.). (2000). *Of minds and molecules: New philosophical perspectives on chemistry*. Oxford, UK: Oxford University Press.

BouJaoude, S., Dagher, Z., & Refai, S. (2017). The portrayal of nature of science in Lebanese ninth grade science textbooks. In C. V. McDonald & F. Abd-El Khalick (Eds.), *Representations of nature of science in school science textbooks: A global perspective* (pp. 79–97). London: Routledge.

Brandon, R. (1994). Theory and experiment in evolutionary biology. *Synthese, 99*, 59–73.

Brock, W. H. (1992). *The Fontana history of chemistry*. London: Fontana Press.

Bruner, J. (1966). *Toward a theory of instruction*. Cambridge, MA: Harvard University Press.

Cartwright, N. (1983). *How the laws of physics lie*. Oxford, UK: Oxford University Press.

Christie, J. R., & Christie, M. (2003). Chemical laws and theories: A response to Vihalemm. *Foundations of Chemistry, 5*(2), 165–174.

Cleland, C. E. (2001). Historical science, experimental science, and the scientific method. *Geology, 29*(11), 987–990.

Corlett, J. A. (1991). Some connections between epistemology and cognitive psychology. *New Ideas in Psychology, 9*(3), 285–306.

Downes, S. M. (1993). The importance of models in theorizing: A deflationary semantic view. *Philosophy of Science Association, 1*, 142–153.

Duschl, R., & Erduran, S. (1996). Modeling the growth of scientific knowledge. In G. Welford, J. Osborne, & P. Scott (Eds.), *Research in science education in Europe: Current issues and themes* (pp. 153–165). London: Falmer Press.

Duschl, R. A. (1990). *Restructuring science education: The importance of theories and their development*. New York: Teacher' College Press.

Dutch, S. I. (1982). Notes on the fringe of science. *Journal of Geological Education, 30*, 6–13.

Erduran, S. (1999). *Merging curriculum design with chemical epistemology: A case of teaching and learning chemistry through modeling*. Unpublished PhD dissertation, Vanderbilt University, Nashville.

Erduran, S. (2014). Beyond nature of science: The case for reconceptualising 'science' for science education. *Science Education International, 25*(1), 93–111.

Erduran, S., & Dagher, Z. R. (2014). *Reconceptualizing the nature of science for science education: Scientific knowledge, practices and other family categories*. Dordrecht, The Netherlands: Springer.

Erduran, S., Kaya, E., & Dagher, Z. R. (2018). From lists in pieces to coherent wholes: Nature of science, scientific practices, and science teacher education. In J. Yeo, T. Teo, & K. S. Tang (Eds.), *Science education research and practice in Asia-Pacific and beyond* (pp. 3–24). Singapore: Springer.

Erduran, S., Cullinane, A., & Wooding, S. (2019). Assessment of practical chemistry in England: An analysis of methods assessed in high stakes examinations. In M. Schultz, S. Schmid, & G. Lawrie (Eds.), *Research and practice in chemistry education: Advances from the 25th IUPAC international conference on chemistry education 2018*. Dordrecht: Springer.

Giere, R. (1991). *Understanding scientific reasoning* (3rd ed.). Fort Worth, TX: Holt, Rinehart, and Winston.

Grosslight, K., Under, C., Jay, E., & Smith, C. (1991). Understanding models and their use in science: Conceptions of middle and high school students and experts. *Journal of Research in Science Teaching, 29*, 799–822.

Hjorland, B., Scerri, E., & Dupre, J. (2011). Forum: The philosophy of classification. *Knowledge Organization, 38*(1), 9–24.

Hoffman, R., Minkin, V. I., & Carpenter, B. K. (1997). Ockham's razor and chemistry. *Hyle – An International Journal for the Philosophy of Chemistry, 3*, 3–28.

Irzik, G., & Nola, R. (2011). A family resemblance approach to the nature of science. *Science & Education, 20*, 591–607.

Justi, R., & Gilbert, J. K. (2000). History and philosophy of science through models: Some challenges in the case of 'the atom'. *International Journal of Science Education, 22*(9), 993–1009.

Kaya, E., Erduran, S., Aksoz, B., & Akgun, S. (2019). Reconceptualised family resemblance approach to nature of science in pre-service science teacher education. *International Journal of Science Education, 41*(1), 21–47.

Kuhn, T. (1962/1970). *The structure of scientific revolutions* (2nd ed.). Chicago: University of Chicago Press.

Kwasnik, B. H. (1999). The role of classification in knowledge representation and discovery. *Library Trends, 48*(1), 22–47.

Latour, B., & Woolgar, S. (1979). *Laboratory life: The social construction of scientific facts*. London: Sage.

Leicester, H. M. (1971). *The historical background of chemistry*. New York: Dover Publications.

Leicester, H. M. (1981). Bessel. In C. Gillespie (Ed.), *Dictionary of scientific biography* (Vol. 2, p. 94). New York: Charles Scribner's Sons.

Loughran, J. J. (2006). *Developing a pedagogy of teacher education: Understanding teaching and learning about teaching*. London: Routledge.

Machamer, P., & Woody, A. (1992). A model of intelligibility in science: Using the balance as a model for understanding the motion of bodies. In *Proceedings of the second international conference on the history and philosophy of science and science teaching* (pp. 95–111). Kingston, ON: Queen's University.

McComas, W. F. (1998). The principal elements of the nature of science: Dispelling the myths. In W. F. McComas (Ed.), *The nature of science in science education* (pp. 53–70). Dordrecht, The Netherlands: Springer.

McDonald, C. V. (2016). Evaluating junior secondary science textbook usage in Australian schools. *Research in Science Education, 46*(4), 481–509.

McDonald, C. V. (2017). Exploring representations of nature of science in Australian junior secondary school science textbooks: A case study of genetics. In C. V. McDonald & F. Abd-El Khalick (Eds.), *Representations of nature of science in school science textbooks: A global perspective* (pp. 98–117). London: Routledge.

Radder, H. (2009). The philosophy of scientific experimentation: A review. *Automated Experimentation, 1*(2), 1–8. https://doi.org/10.1186/1759-4499-1-2.

Redhead, M. L. G. (1980). Models in physics. *British Journal for the Philosophy of Science, 31*, 145.

Rice, F. O., & Teller, E. (1938). The role of free radicals in elementary organic reactions. *Journal of Chemical Physics, 6*, 489–496.

Saribas, D., & Ceyhan, G. D. (2015). Learning to teach scientific practices: Pedagogical decisions and reflections during a course for pre-service science teachers. *International Journal of STEM Education, 2*(1), 7.

Scerri, E. R. (2006). Normative and descriptive philosophy of science and the role of chemistry. In D. Baird, E. R. Scerri, & L. McIntyre (Eds.), *Philosophy of chemistry: Synthesis of a new discipline* (pp. 119–128). Dordrecht, The Netherlands: Springer.

Scerri, E. R. (2007). *The periodic table: Its story and its significance.* New York: Oxford University Press.

Schummer, J. (2014). Aesthetic values in chemistry. *Rendiconti Lincei. Scienze Fisiche e Naturali, 25*, 317–325.

Shapin, S., & Schaffer, S. (1985). *Leviathan and the air pump: Hobbes, Boyle, and the experimental life: Including a translation of Thomas Hobbes, Dialogus Physicus de Natura Aeris.* Princeton, NJ: Princeton University Press.

Suckling, C. J., Suckling, K. E., & Suckling, C. W. (1978). *Chemistry through models.* Cambridge, UK: Cambridge University Press.

Tomasi, J. (1988). Models and modeling in theoretical chemistry. *Journal of Molecular Structure (THEOCHEM), 48*, 273–292.

Trindle, C. (1984). The hierarchy of models in chemistry. *Croatica Chemica Acta, 57*, 1231.

van Keulen, H. (1995). *Making sense: Simulation-of-research in organic chemistry education.* Utrecht, The Netherlands: CD- Press.

Woody, A. (1995). The explanatory power of our models: A philosophical analysis with some implications for science education. In F. Finley, D. Allchin, D. Rhees, & S. Fifield (Eds.), *Proceedings of the third international history, philosophy, and science teaching conference* (Vol. 2, pp. 1295–1304). Minneapolis, MN: University of Minnesota.

# Chapter 3
# Epistemic Beliefs and Teacher Education

## 3.1 Introduction

In a study about pre-service chemistry teachers' beliefs in and knowledge of chemistry, Veal (2004) reported that a pre-service teacher was bothered by how a more experienced teacher presented chemistry concepts to students. He was particularly concerned about how chemistry was presented as if it is magical:

> All right, he's making them the saturated solutions and it says that the solid particles were just buzzing around there at fast speeds until the solute molecules are no longer solid, but in solution. And ... all right, it kind of makes it sound like it's magic. And, I guess it maybe it just bothers me a little bit. Because it's not really magic ... And then he does that again, because he says when he puts the beaker into the water, he says that, it has so much volume that it just turned to solid. And that kind of makes it sound like it's magic, too. Which I read another paper on what makes chemistry hard is the level of information. That's part of why students don't like chemistry is because it's not very real to them, things just kind of happen. (p. 339)

It is indeed concerning if students are not offered any justification of the claims made in chemistry lessons through evidence that substantiate why one should believe in such claims. As in the scenario described above, often in chemistry lessons, students are expected to take the teacher's word as facts, and they may not relate to how and why they should believe in knowledge presented by the teacher. Teaching that deals with knowledge in its final form, delivered from the teacher to the students, immerses students in a mode of unsubstantiated thinking and unquestioned dogma. In contrast, imagine a lesson where the students actually understand what a claim is, why it should be believed in and how it is justified with evidence and reasons. In such a lesson, students would be engaged in producing and evaluating claims. They would question what counts as evidence and how to distinguish reliable and relevant evidence. How, then, would students from either lesson compare in their understanding of how knowledge works in chemistry? Which lesson would you expect to foster understanding that chemistry is not about magic but

© Springer Nature Switzerland AG 2019
S. Erduran, E. Kaya, *Transforming Teacher Education Through the Epistemic Core of Chemistry*, Science: Philosophy, History and Education,
https://doi.org/10.1007/978-3-030-15326-7_3

rather it is about justification of knowledge claims? Imagine also a contrast of the teachers' approaches to these lessons. In the first scenario, the teacher is simply delivering ready-made knowledge, while in the second scenario, the teacher is engaging the students in the evaluation of knowledge. Which teacher would you expect to have good understanding of what knowledge is in chemistry and how it develops? Which teacher would you expect to have metacognitive awareness of how and why we know? Bringing the epistemic core of chemistry (see Chap. 2) to the foreground of teacher education is likely to support teachers in learning how to teach the second scenario. This is because the epistemic core explicitly addresses the how and why questions about chemistry knowledge. Furthermore, it highlights why particular knowledge is pursued in chemistry in the first place (i.e. aims and values) and how it gets produced (i.e. practice, methods) and what it involves (i.e. knowledge).

Arguments about the importance of teachers' understanding of the epistemic dimensions of their subject are not new. Almost half a century ago, Schwab (1962) had argued that expertise in teaching requires both knowledge of content of a domain and knowledge about the epistemology of that domain. Teachers develop the necessary capability of transforming subject into teachable content only when they know how the disciplinary knowledge is structured. For example, a chemistry teacher who understands why and how models contribute to chemical knowledge will be better able to facilitate students' understanding of the role of models in chemistry. Numerous studies (e.g. Lampert, 1990; Rollnick, Bennett, Rhemtula, Dharsey, & Ndlovu, 2008; Shulman, 1987) have illustrated the centrality of disciplinary knowledge in good teaching. The challenge facing teacher education is that teachers in general have had little exposure to epistemic issues in relation to chemical knowledge beyond subject knowledge which is primarily about concepts such as chemical equilibrium, bonding and gas laws. In other words, teachers are typically indoctrinated into the conceptual outcomes of chemistry without much engagement with the rationale nor the processes that underpin their development. Erduran, Aduriz-Bravo and Mamlok-Naaman (2007) argued that philosophy of chemistry can potentially provide some tools for the epistemic empowerment of chemistry teachers. For example, they discussed how the philosophical theme of reductionism can be incorporated in teacher education. Erduran and colleagues' advancement of the use of philosophy of chemistry in chemistry teacher education is underpinned by a wider consensus on the benefits of including history and philosophy of science (HPS) in science education (see Chap. 1).

As a broad field, HPS offers many implications for teacher education. The content selected for teacher education could be related to some epistemological and ontological themes such as rationalism and realism (Giere, 1999). Teacher education could also focus on particular epistemic aspects of science such as the nature of models (e.g. Giere, 1999; Izquierdo & Aduriz-Bravo, 2003) and theories (e.g. Duschl, 1990). Epistemology in relation to science teacher education can be considered in numerous other senses. For example, the very epistemic aims of education can be questioned. Another sense concerns the teachers' and students' own epistemologies in relation to what scientific knowledge is and how and why it comes to be. In this chapter, we illustrate the relevance of epistemology in teacher education and explore related concepts such as "epistemic beliefs", "epistemic cognition" and "personal epistemologies". Furthermore, research evidence on teachers' knowledge and learning is reviewed to provide an indication of issues in the broader literature. The chapter concludes with some potential strategies such as argumentation (e.g. Erduran & Jimenez-Aleixandre, 2007), visualisation (e.g. Gilbert, Reiner, & Nakhleh, 2008) and analogies (e.g. Aubusson, Treagust, & Harrison, 2009) that can be used in teacher preparation programmes to support the development of teachers' epistemic thinking through the epistemic core (see Chap. 2). Although these strategies are not exhaustive, they provide focus about a set of strategies that can be used in teacher education to teach some fairly abstract learning outcomes such as the learning of epistemic themes. Through these and similar strategies, the epistemic core can be made accessible (e.g. through analogies); unpacked, evaluated and discussed (e.g. through argumentation); and shared and monitored (e.g. through visualisation) in teacher education.

## 3.2 Epistemology and Teacher Education

Broadly speaking epistemology, or theory of knowledge, plays a significant role in the development of educational theory and practice. This sense of application of epistemology in teacher education points to the epistemic aims of education. For example, Ortwein, McCullough and Thompson (2015) investigated how educators understand knowledge within their practice. These researchers questioned the extent to which educational theories have actually shaped teachers' attitudes towards knowledge and teaching practices. They explored two related questions: (a) to what extent does *theoretical knowledge* affect teachers' understandings of knowledge and its function in education and (b) what experiential factors shape teachers' understandings of knowledge and its role within educational practice? Ortwein and colleagues concluded that teachers' philosophical understanding of knowledge when applied to education is difficult for teachers to elucidate. Experience combined with indirect theoretical influences shape teachers' understandings of the function of knowledge and related educational epistemology.

Research about the relationship between teachers' scientific epistemological views (SEVs) and science instruction is a different sense of how epistemology can be considered in teacher education. Tsai (2007) worked with science teachers and their students, and conducted classroom observations to examine any coherences between teachers' SEVs and their (a) teaching beliefs, (b) instructional practices, (c) students' SEVs and (d) students' perceptions towards actual science learning environments. The findings suggested adequate coherences between teachers' SEVs and their teaching beliefs as well as instructional practices. The teachers with relatively positivist-aligned SEVs tended to draw attention to students' science scores in tests and allocate more instructional time on teacher-directed lectures, tutorial problem practices or in-class examinations, implying a more passive or rote perspective about learning science. In contrast, teachers with constructivist-oriented SEVs tended to focus on student understanding and application of scientific concepts, and they adopted more time on student inquiry activities or interactive discussion. These findings are quite consistent with the results about the coherence between teachers' SEVs and students' perceptions towards science learning environments, suggesting that the constructivist-oriented SEVs appeared to foster the creation of more constructivist-oriented science learning environments. Finally, although this study provided some evidence that teachers' SEVs were likely related to their students' SEVs, the teachers' SEVs and those of their students were not coherent.

There is some research evidence on how epistemological ideas may be directly taught in teacher education. Sendur, Polat and Kazanci (2017) investigated the impact of a course on history and philosophy of chemistry on prospective chemistry teachers' perceptions of chemistry and the chemist. The study was conducted with 38 prospective chemistry teachers. A creative comparison questionnaire and semi-structured interviews were used as data collection instruments in the study. The questionnaire was administered to the prospective teachers in the form of a pre-test, post-test and retention test. Results of the analysis showed that the prospective teachers produced creative comparisons related to chemistry in the pre-test that mostly relied on their own experiences and observations but that in the post-test and retention test, their comparisons mostly contained references to the role of chemistry in daily life, its development and its facilitating aspects. Similarly, it was observed that in the pre-test, the prospective teachers made creative comparisons regarding the chemist that related mostly to the laboratory but that the post-test and retention test rather contained the aspects of chemists as researchers, meticulous persons, facilitators and managers. The results of the interviews indicated that a large majority of the prospective teachers were able to fully reflect on the inadequacy of their prior knowledge about chemistry.

A similar direct approach on the teaching of epistemology to pre-service teachers was utilised by Veli-Matti Vesterinen (2012) who reported an educational design research project on nature of science (NOS) instruction. Educational design research is the systematic study of the design and development of educational interventions for addressing complex educational problems. It advances the knowledge about the characteristics of designed interventions and the processes of design and development. The project consisted of four interconnected studies and documents two iterative design research cycles of problem analysis, design, implementation and evaluation. The first two studies described how NOS is presented in the national frame curricula and upper secondary school chemistry textbooks. These studies provided a quantitative method for analysis of representations of NOS in chemistry textbooks and curricula, and described the components of domain-specific NOS for chemistry education. The other two studies documented the design, development and evaluation of the goals and instructional practices used on the course. Four design solutions were produced: (a) description of central dimensions of domain-specific NOS for chemistry education, (b) research group visits to prevent the diluting of relevance to science content and research, (c) a teaching cycle for explicit and structured opportunities for reflection and discussion and (d) collaborative design assignments for translating NOS understanding into classroom practice. The evaluations of the practicality and effectiveness of the design solutions were based on the reflective essays and interviews of the pre-service teachers, which were collected during the course, as well as on the four in-depth interviews of selected participants, collected a year after they had graduated as qualified teachers. The results suggest that one critical factor influencing pre-service chemistry teachers' commitment to teach NOS was the possibility to implement NOS instruction during the course. Thus, the use of collaborative peer teaching and integrating student teaching on NOS instruction courses was suggested as a strategy to support the development of the attitudes, beliefs and skills necessary for teaching NOS. In summary, infusing epistemological perspectives in chemistry education is a challenge for teachers who have had little exposure to issues of chemical knowledge from a meta-perspective (Erduran et al., 2007). Teachers' epistemic beliefs about knowledge and knowing are important to consider if the teacher education is going to target the learning of epistemic thinking.

## 3.3  Epistemic Beliefs

Research on teachers' epistemic beliefs have focused on a range of related themes such as personal epistemologies and epistemic cognition. These lines of research have often drawn on Perry's (1970) research on the intellectual development of students in Harvard as they progressed through their degree programme and university experiences. As Hofer (2016) notes:

> Although Perry did not use the term epistemological development, researchers who followed recognized in his stages an evolving understanding of what it means to know and, accordingly, how one goes about the process of learning and understanding. (p. 23)

Perry (1970) noted nine positions that progress stagelike fashion. Positions 1 and 2 can be clustered and termed "dualism". Learners at this stage of development would be characterised by a dualistic, absolutist, right-and-wrong view of knowledge. They would consider authorities to be the source of knowledge. These authorities know the truth and transmit it to the learner. Positions 3 and 4, collectively termed "multiplicity", are indicated by successive modification of the right-vs-wrong duality. There is a recognition in diversity of perspectives and that a third option exists other than right or wrong: "not yet known". In Position 3, this third option is considered "temporarily unknown" and will become known in the future. Position 4 might argue that we will never know for sure, and so we must rely on our own thinking. This leads to a stance that there is no nonarbitrary basis for determining what is right, and so all views are equally valid. Position 5 is the beginning of "contextual relativism" and is the significant pivot point of the developmental trajectory where individuals shift from a form of dualistic perspective to a relativistic one. It differs from the sort of pseudo-relativism of Position 4 in that there is now a self-conscious awareness as the individual as an active maker of meaning. At this stage, knowledge is considered context-bound, with only a small number of right/wrong dichotomies. Positions 6–9 are labelled together as "commitment within relativism". The distinctions between these final stages are described as being more qualitative than structural, and although Perry proposed them as part of the developmental trajectory, they were not commonly found among college students. Development through these positions involves a shift from the intellectual to the ethical, and development of one's identity in a contextually relativistic world.

Subsequent research on epistemological beliefs included Schommer's work (1990) where a quantitative approach was pursued including large-scale, paper-and-pencil assessments of personal epistemology. Schommer's paper-and-pencil measures have been influential on more recent work (e.g. Lunn Brownlee, Schraw, & Berthelsen, 2011). Schommer conceptualised personal epistemology as a system of independent dimensions of beliefs. She proposed five dimensions: structure, certainty, source of knowledge and control and speed of acquisition of knowledge. An individual could hold more advanced beliefs in one dimension while simultaneously holding less advanced views in another dimension. While Schommer's model was perceived as providing a nuanced approach to the categorisation of beliefs, it also faced much criticism. For example, Hofer and Pintrich (1997) argued that the inclusion of learning dimensions was problematic. The dimensions of "fixed learning ability" and "quick learning" were arguably not of an epistemological or philosophical nature. Although these issues of intelligence or ability could be correlated with epistemological issues, the authors argued that they were fundamentally different constructs that should remain separate. Additionally, they critiqued Schommer's instrument indicating that it contained "vague remnants of personality measurement with questionable relevance" (Hofer & Pintrich, 1997, p. 109).

Hofer and Pintrich (1997) made a number of propositions for further research and developed a new theoretical model of personal epistemology. They highlighted the difficulty in naming and defining the construct based on the existing research "to the extent that it is sometimes unclear to what degree researchers are discussing the same intellectual territory" (p. 111). There are some key distinctions between the proposed model from Hofer and Pintrich compared to early counterparts. They propose that personal epistemology is more than a collection of independent beliefs, but rather they are "personal theories" about the nature of knowledge and processes of thinking (p. 117). They also proposed that these theories can be both domain-general and domain-specific, rather than necessarily one or the other. They suggest that it is a false dichotomy and that these general and specific beliefs operate in a complementary and interactive way. Hofer and Pintrich (1997, p. 120) proposed four dimensions as follows:

- *Certainty of knowledge* – The extent to which knowledge is viewed as fixed or fluid. At lower levels of development, an individual may consider knowledge to be existing with certainty. In this case, knowledge cannot be doubted, and all experts would come up with the same answer to a question in the field. That answer would not be likely to change over time. Alternatively, at higher levels of development, individuals would be open to the idea that theories are modified over time as more information is gathered and that knowledge is not certain or absolute.

- *Simplicity of knowledge* – In this dimension the continuum ranges from knowledge as a set or accumulation of facts to knowledge as highly interrelated concepts. At lower levels of development, answers are simple and straightforward; knowledge is discrete, concrete and easily knowable. At higher levels of development, the meaning of a concept is understood as complex, relative to other concepts, and dependant on context.
- *Source of knowledge* – At lower levels, knowledge resides outside the self. Knowledge is transmitted from expert/authority to the learner and should not be questioned. Further along in development, individuals might consider knowledge as being constructed in interaction with others that the person would move from being a spectator to being an active participant in constructing knowledge, using logic and evidence.
- *Justification for knowing* – This dimension is concerned with how individuals evaluate knowledge and how they use or evaluate evidence, authority and expertise. As individuals learn to evaluate evidence and to substantiate and justify their beliefs, they move through a continuum of dualistic beliefs to the multiplistic acceptance of opinions to reasoned justification for their beliefs.

The model as proposed by Hofer and Pintrich (1997) and subsequent empirical accounts based on the adaptations of their framework have made a significant impact in the literature (e.g. Buehl & Alexander, 2002). However, this dimensional model and those proposed by others like Schommer (1990) have also attracted criticism. For example, Chinn and Rinehart (2016) objected to the polarised view that beliefs in uncertainty and tentativeness are sophisticated while beliefs in certainty of knowledge are unsophisticated. They argued for a more nuanced perspective where there is acknowledgement various elements of science have different levels of certainty, ranging from "core" elements known with a high degree of certainty to more "peripheral" elements which are less certain.

The research on epistemic beliefs has been considered in the context of teachers' beliefs and applied to teacher education. Bondy et al. (2007) used personal epistemology a framework to example how pre-service teachers progress in their learning. They described how the pre-service teachers' epistemic beliefs influence their engagement with knowledge in their teacher education programmes, by influencing their aims and expectations. Joram (2007) conducted a comparative study focusing on professors of education, in-service teachers and pre-service teachers. Her research highlighted the differences of epistemic beliefs between the three groups and showed the apparent influence on their attitude towards educational research.

Practicing teachers' epistemic cognition has been shown to influence their teaching approach, the strategies they employ in the classroom and their expectations for students (Lunn Brownlee et al., 2011). Pre-service teachers' epistemic cognition is thought to impact the depth of understanding achieved during teacher education courses and teachers' planning and assessment in subsequent practice (Buehl & Fives, 2016). Yadav, Herron and Samarapungavan (2011) considered the ways in which pre-service teachers' epistemic cognition was important for teacher preparation. The authors argued that pre-service teachers' epistemic cognition played a role in their perceptions and attention when observing other teachers and in the teaching goals that they developed. Regarding pre-service teachers' epistemic cognition and teaching goals, Kang (2008) reported that pre-service science teachers who viewed science knowledge as consisting of facts set the goal of having students utilise science knowledge, whereas those who viewed science knowledge as evolving in nature were more likely to aim to have students develop thinking skills necessary to conduct scientific inquiry.

## 3.4   Teachers' Knowledge and Learning

Although epistemic beliefs are important in relation to teachers' learning about the epistemology of their subject domain, there are broader sets of issues to consider in teacher education. There is considerable research on science teachers' learning and professional development more generally (e.g. Kind, 2009; Minstrell & VanZee, 2000; Simon, Osborne, & Erduran, 2003; Supovitz & Turner, 2000). Some professional development strategies rely on theory-driven initiatives. For example, much research on argumentation in science teacher education has relied on the theoretical definition of argument based on Toulmin's (1958) conceptualisation of argument (e.g. Erduran & Jimenez-Aleixandre, 2007). Teacher education initiatives have highlighted factors that impact teacher change (e.g. Fullan, 2001). Early approaches to teacher learning had little sustained impact and were underpinned by beliefs that teacher learning is a linear process and that educational change is a "natural consequence of receiving well-written and comprehensive instructional materials" (Hoban, 2002, p. 13). A more complex view of professional development is required, incorporating professional learning systems that only bring about sustained change over a long period of time. Educational change is complex and takes time (Fullan, 2001), and it was never anticipated that fundamental and substantial changes could be achieved within the time scale of 1 year. However, within the context of accountability and high stakes assessment that pervade many educational systems around the world, teacher educators strive to have impact within the existing constraints.

Researchers have taken note of the differentiation in the issues related to beginning and in-service teachers (e.g. Cochran, DeRuiter, & King, 1993; Friedrichsen, Van Driel, & Abell, 2010). It is widely noted that experienced teachers can benefit from professional development input, while novice teachers are often caught up on classroom management issues with limited ability to apply innovative strategies in their lessons (e.g. Luft et al., 2011). However, when experienced science teachers change schools or have to teach unfamiliar content, they resume the role of the novice science teachers again (Loughran, 2007). Teachers' conceptions of nature of science are also known to be important in shaping their teaching, and they have been shown to be naïve conceptions of NOS, similar to those of students (e.g. Akerson, Morrison, & Roth McDuffie, 2006; Schwartz & Lederman, 2002).

Shulman (1986) provided a powerful construction, "pedagogical content knowledge" (PCK), to illustrate this kind of understanding and knowledge that teachers need to have. He described PCK as "The most useful forms of content representation... the most powerful analogies, illustrations, examples, explanations, and demonstrations— in a word, the ways of representing and formulating the subject that makes it comprehensible for others" (p. 9). Grossman (1990) added two other components to Shulman's original PCK components, that is, knowledge of curriculum and knowledge of purposes for teaching. Another elaboration that has been very influential in the context of science education is the model of Magnusson, Krajcik and Borko (1999). This model added three components to the original ones of Shulman: orientation to teaching science (i.e. knowledge and beliefs about purposes and goals for teaching), knowledge of science curricula and knowledge of assessment of scientific literacy. A more recent perspective on teacher knowledge uses a transformative, yet structured model of teacher professional knowledge and skills. This model proposed by Gess-Newsom (2015) incorporates ideas from Shulman's (1987) such as PCK as well as other concepts such as teacher professional knowledge bases (TPKBs) and topic-specific professional knowledge (TSPK). The model makes explicit that content for teaching occurs at the topic levels (i.e. atom and chemical equilibrium) and not at the disciplinary level (i.e. chemistry or science). Furthermore, there's problematisation of how subject matter, pedagogy and context can be considered in unison. It recognises the difference between public knowledge and knowledge held by a professional. In order to impact teachers' learning, teacher educators need to construct practical and simple ideas that would have pedagogical value to help teachers construct the knowledge they need for understanding the purpose of teaching chemistry from an epistemic perspective. Moreover, the role of the teacher in promoting reflections on the nature of chemistry including the epistemic core of chemistry (i.e. epistemic aims, values, methods, practice) would mean a shift away from the role of an authority figure providing right answers (Bay, Reys & Reys, 1999) to one where the teacher is a facilitator. Teacher educators need to be mindful of providing the opportunity to enable teachers to meet and work collaboratively. In doing so, teachers are provided with a forum for deliberation (Spillane, 1999).

Teaching of abstract notions like the epistemic core of chemistry demands metacognitive awareness. The literature on metacognition is vast (e.g. Peters & Kitsantas, 2010). As a major contributor to this literature, Flavell (1979) categorised metacognition into metacognitive knowledge and metacognitive experiences. Metacognitive knowledge involves knowledge and beliefs about the variables influencing the courses and outcomes of cognitive processes. Flavell discussed the metacognitive knowledge in three categories which were labelled as person, task and strategy variables. The person variable refers to one's belief about the nature of oneself and other people as cognitive processors. For example, someone can believe that she/he learns something better by taking notes than by just listening or she/he believes that her/his friend is more competent in electricity than in mechanics. Task variable includes knowledge and judgements about task goals and demands such as an individual's conceptualisation of the task in terms of whether a specific task is easy or hard to achieve or knowledge about the sources (cognitive as well as outside help) required to accomplish a task. The strategy variable refers to knowledge about strategies useful for certain goals or sub-goals as well as knowing the necessary strategies for specific cognitive tasks. Schraw and Moshman (1995) considered metacognition as knowledge about cognition which refers to one's knowledge about her/his own cognition. It consists of three subcomponents: (a) declarative, (b) procedural and (c) conditional knowledge. Declarative knowledge is defined as one's knowledge about oneself as a cognitive processor. Procedural knowledge involves knowledge about execution of procedures for a specific cognitive task. The conditional knowledge refers to knowledge of why and when to use a particular strategy for a particular cognitive task. More contemporary accounts of metacognition research in the context of science education include Zohar and team's work on the role of metacognition in scientific inquiry. Zohar (2012) discusses meta-strategic knowledge or MSK as a subcomponent of metacognition. MSK is the "thinking behind the thinking" (meta-level of thinking) rather than the "thinking behind the doing" (Zohar & Ben-David, 2008).

Despite particular limitations in pre-service teachers' skills, providing them with high-quality topic specific professional development is shown to influence practice (Smylie, 1989). Supovitz and Turner (2000, p. 964) identified the following critical features of high-quality professional development: (a) immerse participants in inquiry, questioning and experimentation; (b) be intensive and sustained; (c) engage teachers in concrete teaching tasks and be based on teachers' experiences with students; (d) focus on subject-matter knowledge and deepen teachers' content skills; (e) be grounded in a common set of professional development standards and show teachers how to connect their work to specific standards; and (f) be connected to other aspects of school change. Apart from an understanding of the content (or subject) domain such as chemistry and the epistemology of the domain, teachers need to have understanding of how to transform these notions into teachable scenarios (Loucks-Horsley et al. 1990). Similarly, a common vision of effective professional development was proposed by Loucks-Horsley, Hewson, Love, and Stiles (1998) and Loucks-Horsley et al. (1990). According to that shared vision, the best professional development experiences for science educators include the following guidelines:

- They are driven by a clear, well-defined image of effective classroom learning and teaching.
- They provide teachers with opportunities to develop knowledge and skills and broaden their teaching approaches, so they can create better learning opportunities for students.
- They use instructional methods to promote learning for adults which mirror the methods to be used with students.
- They build or strengthen the learning community of science and mathematics teachers.
- They prepare and support teachers to serve in leadership roles if they are inclined to do so. As teachers master the skills of their profession, they need to be encouraged to step beyond their classrooms and play roles in the development of the whole school and beyond.

- They consciously provide links to other parts of the educational system.
- They include continuous assessment.

How then can pre-service teachers' learning of the epistemic core of chemistry be supported? What strategies can teacher educators use to promote and support pre-service teachers' understanding of epistemic aspects of chemistry? The inclusion of the epistemic core of chemistry in chemistry teacher education demands the teaching and learning of abstract ideas that are often unfamiliar to pre-service teachers. Researchers have used approaches that vary along a continuum from implicit to explicit in their attempts to enhance students and teachers' NOS views (e.g. Abd-El-Khalick & Lederman, 2000). Grandy and Duschl (2007) have disputed such distinctions on the basis that they "greatly oversimplify the nature of observation and theory and almost entirely ignores the role of models in the conceptual structure of science" (p. 144).

The *explicit-reflective approach* (e.g. Akerson, Abd-El-Khalick & Lederman, 2000) that has been argued to develop students' NOS views is relatively more effective than an *implicit approach* (e.g. Haukoos & Penick, 1985) that utilises hands-on or inquiry science activities lacking explicit references to NOS. The implicit approach assumes that learners will learn NOS as a natural consequence of the engagement in scientific inquiry activities (Schwartz, Lederman & Crawford, 2004). However, research on students, teachers and scientists' views of NOS show that the subject who engages in scientific inquiry alone does not necessarily develop contemporary views of NOS (Bell, Blair, Lederman & Crawford, 2003). The main criticism to the implicit approach is that during teaching, teachers and students are often unclear about the learning aims of teaching activities related to NOS (Leach, Hind, & Ryder, 2003). In most cases, science is not there to be discovered through close scrutiny of the natural world; rather, it involves introducing students to the scientific ideas and ideas about NOS (Leach & Scott, 2002). Scott, Leach, Hind, and Lewis (2006) have drawn attention to the fact that "learning science involves being introduced to the language of the scientific community and this can be achieved through the agent of a teacher or some other knowledgeable figure" (p. 62). Similar arguments can be extended to the learning of teachers themselves. Given future teachers will be expected to take on the role of interpreters and mediators of the language of the scientific community (Mortimer & Scott, 2003), they themselves need to learn about these modes of thinking and behaving. In this sense, teacher education needs to model to pre-service teachers what has been shown to be effective through research.

A relevant concept in teachers' knowledge and skills concerns "epistemic cognition". Epistemic cognition typically refers to beliefs, epistemic development, epistemological beliefs and personal epistemologies (Greene, Azevedo, & Torney-Purta, 2008). Since its inception in the 1970s, research on epistemic cognition has experienced substantial growth (Greene, Sandoval, & Braten, 2016). Earlier described as layperson's folk epistemologies, unexamined understandings or commonsense theories and "untutored views about the nature of knowledge" (Kitchener, 2002, p. 89), the use of the term epistemic cognition now reflects "how people acquire, understand, justify, change, and use knowledge in formal and informal contexts" (Greene et al., 2016, p. 1). Chinn and colleagues' definition is directed at cognitions about a network of interrelated epistemic topics including knowledge, its sources and justification, belief, evidence, truth, understanding and explanation (Chinn, Buckland, & Samarapungavan, 2011). Researchers have argued that one way of enriching educational research on epistemic cognition is to take a closer look at philosophical literature (Hofer, 2016), which has several implications for expansion of the construct (Chinn et al., 2011). Building on an extensive review of educational and philosophical literature (Chinn et al., 2011), Chinn and colleagues developed the AIR framework with a specific focus on epistemic aims, ideals and reliable processes (Chinn & Rinehart, 2016). The first component, epistemic aims and values, draws attention to the idea that people can have different epistemic aims other than ascertaining knowledge, such as developing true beliefs, understanding or wisdom. As such, epistemic cognition widens the scope of the use of the term epistemic (Chinn & Rinehart, 2016).

Although epistemic themes have been targeted for pre-service teacher education and conceptual frameworks have been proposed to facilitate their realisation, there is also evidence that beginning science teachers face numerous challenges (Luft et al., 2011). For example, when beginning chemistry teachers are given innovative pedagogical materials, they continue to teach in a traditional manner (Roehrig & Luft, 2004). Duschl and Gitomer (1997) observed that teachers tend to use curriculum materials on the basis of practicality rather than choosing them based on how they promote conceptual understanding and scientific reasoning. The difficulties with beginning teachers are compounded by the fact that many teachers drop out of teaching soon after they qualify to teach. For example, in the USA, Ingersoll (2003) found that approximately 50% of teachers leave within 5 years of their teaching careers. While the systemic problems are beyond the scope of our undertaking in this book, we contend that it is plausible that teacher education interventions enthuse pre-service teachers about their subject domain, and that teacher education can capitalise on some strategies that facilitate pre-service teachers' meaningful learning such that they are motivated to remain in the profession.

## 3.5 Strategies for Supporting Chemistry Teacher's Epistemic Thinking

Lunn-Brownlee and Schraw (2016) identified two main ways in which changes to epistemic beliefs might take place: as a result of engagement in higher order thinking and explicit reflection on epistemic beliefs. Engagement in either aspect needs to be structured and supported. There are numerous strategies that can be considered for inclusion in teacher education to support pre-service teachers' learning of the epistemic core of chemistry. Here we will provide some relevant examples such as argumentation (e.g. Erduran & Jimenez-Aleixandre, 2007), visualisation (e.g.

Eilam & Gilbert, 2014) and analogies (e.g. McComas, 2014). These strategies are implicit in the visual representations used to capture the epistemic core in chemistry as discussed in Chap. 2. For example, the Benzene Ring Heuristic of scientific practices as defined by Erduran and Dagher (2014) articulates how scientists use data originating from the real world to generate models, explanations and predictions. The modelling practice is inclusive of analogies as well as argumentation where alternative claims may be debated about the structure and function of matter. BRH itself is a visual representation that is meant to provide a simple yet inclusive account of scientific practices. As the following sections will illustrate, argumentation is consistent with Mortimer and Scott's (2003) recommendation that teachers need to learn the "language" of the scientific community including the modes of reasoning and thinking that underpin that language. The use of visual representations is consistent with the recommendation regarding the "explicit-reflective approach" as argued by Akerson and colleagues (2000) given visual representations make explicit concepts and highlight them in visual form for easier access as compared to a purely text-based account. Finally, the use of analogies can be considered as vehicles in the explicit consideration of particular concepts, processes and relationships that are being developed in pre-service teachers' understanding of the nature of chemistry. In other words, by engaging with familiar concepts and other tools, it is expected that pre-service teachers will begin to make sense of the complex and difficult philosophical considerations that underpin chemistry.

### 3.5.1 Argumentation

Argumentation (i.e. the coordination of evidence and theory to support or refute an explanatory conclusion, model or prediction) has emerged as a significant area of research in science education in the past few decades (Erduran, Ozdem, & Park, 2015). Argumentation can be described as a kind of discourse through which knowledge claims are individually and collaboratively constructed and evaluated in the light of empirical or theoretical evidence (Erduran & Jimenez- Aleixandre, 2007). A recent edited book has highlighted the relevance of argumentation for chemistry education, including its role in learning in the laboratory and specific contexts such as organic and physical chemistry (Erduran, 2019). As a relatively unfamiliar strategy, argumentation needs to be appropriated by children and explicitly taught through suitable instruction, task structuring and modelling (e.g. Mason, 1996). Likewise, teachers' skills in coordinating argumentation-based lessons need to be supported through professional development (e.g. Simon, Erduran & Osborne, 2006).

There are at least three theoretical bodies of research framing argumentation studies: (a) developmental psychology, including the distributed cognition perspective; (b) language sciences, for instance, the theory of communicative action; and (c) science studies, for instance, drawing on history, philosophy and sociology of science. As Erduran and Jimenez-Aleixandre (2007) argued, rather than being a one-way relationship, argumentation studies and science education have the potential to inform these perspectives, leading to fruitful interactions. The study of how argumentation studies have been informed by foundational perspectives is important in setting the scene for potential reciprocal interdisciplinary investigations of argumentation (Erduran & Jimenez-Aleixandre, 2007) with contributions from science education research to other fields. For example, (a) the discussions about to what extent argumentation research in science education contributes to cognitive and metacognitive processes would inform the situated cognition perspective (Brown & Campione, 1990); (b) the development of communicative competences and particularly critical thinking by means of argumentative science education would add to the theory of communicative action (Habermas 1981); and (c) understanding the development of reasoning through argumentation in science education could extend our knowledge about teaching and learning philosophy of science (Giere, 1999) as well as developmental psychology (Kuhn & Crowell, 2011).

There is limited research on teachers' learning and professional development in argumentation (e.g. Simon et al., 2006; Zembal-Saul, 2009; Zembal-Saul & Vaishampayan, 2019; Zohar, 2008). Anat Zohar, who has contributed extensively to research in science teaching and teacher education, states that "...until very recently, very little work has been done specifically about teacher education and professional development in the field of argumentation" (Zohar, 2008, p. 246). A plethora of research studies have focused on the teaching of higher order thinking skills (e.g. Zembal-Saul, 2009), a related skills-set that facilitate the processes of argumentation. However the incorporation of the "epistemic" components of argumentation in the characterisation of argumentation in teaching is still quite challenging for science teachers. Teachers not only need to coordinate the cognitive but also the epistemic goals of science education in developing scientific inquiry. Teachers need to learn about effective pedagogical strategies that would enable students to carry out activities that would rely on the use of evidence-based reasoning, critical thinking and argumentation (Zembal-Saul, 2009; Zohar, 2008).

Learning to teach argumentation is a goal not only for in-service but also pre-service teachers (e.g. Erduran, 2006; Kelly, Druker, & Chen, 1998; Richmond & Striley, 1996; Zembal-Saul, Munford, Crawford, Friedrichsen, & Land, 2002). The models of professional development of in-service teachers rely on the inclusion of mentor teachers in the training of pre-service trainee teachers (e.g. Erduran, 2006; Maloney & Simon, 2006) typically involving both higher education-based training and school-based practical experience. The provision for professional development of in-service teachers tends to be sporadic with few comprehensive trends. In England and Wales, for instance, the Science Learning Centres have been instrumental in the delivery of professional development on the argumentation-related components of the curriculum in a systematic way (Erduran & Jimenez-Aleixandre, 2012).

For argumentation to take place in science classrooms, teachers need to facilitate students to assume an active role in discussions (Jimenez-Aleixandre & Erduran, 2007). Yet, as Newton, Driver and Osborne (1999) indicated, science lessons tend to be conducted with a heavy emphasis on question and answer interactions, and the teacher-dominant practices in science lessons do not tend to involve activities that support discussion, argumentation or the social construction of knowledge.

Therefore, a significant issue about teaching argumentation is the need for teachers to coordinate and mediate related epistemic practices of science in the learning environment so as to allow student participation in discussions (Duschl, 2008; Simon et al., 2006).

## 3.5.2 Visualisation

The role of visualisation in science education has been highlighted extensively (e.g. Dori & Barak, 2001; Gilbert et al., 2008). Research has highlighted the importance of visual representations both for teachers and students. The use of multiple representations in general is an important part of teachers' knowledge of science, and they can play an important role in the explanation of scientific ideas (Eilam & Gilbert, 2014). The research literature in science education is extensive and includes a whole range of emphases. For example, the contribution of visual representations to cognitive understanding in science has been investigated (Gilbert, 2010). There are studies focusing on the role of visualisation in scientific practices themselves (e.g. Evagorou, Erduran & Mantyla, 2015). Studies in science education have explored the use of images in science textbooks (Dimopoulos, Koulaidis, & Sklaveniti, 2003), students' representations or models when doing science (Gilbert et al., 2008; Lehrer & Schauble, 2012) and students' images of science and scientists (Chambers, 1983). Eilam, Poyas, and Hashimshoni (2014) investigated science and mathematics teachers from diverse backgrounds, while they generated visual representations to represent textual data. The findings indicated that teachers had difficulty producing visual representations.

Visual representations exist in two ontological forms (Gilbert, 2005). The first of these is as internal representations which are the personal mental constructions of an individual, typically referred to as mental images. The second is as external representations which are open to inspection by others. Research illustrates that when both visual- and text-based information are considered together, readers process text separately from diagrams and form two representations: one textual and one visual (Mayer, 2005). In Mayer's model, the verbal model and the pictorial model are first fully formed (i.e. both the picture and text are understood), each within a limited capacity working memory system (i.e. auditory working memory and visual working memory). Only after each mental model is formed does the crucial step of integration occur, in which referential connections are formed between the two models and prior knowledge. With regard to the design of diagram comprehension interventions, Mayer's theory calls for text comprehension skills, image comprehension skills, adequate auditory and visual working memory capacity and referential connection skills.

Visual representations enable teachers and students to make connections between their own experience and scientific concepts and therefore gain insight into abstract ideas about science (Post & Cramer, 1989). Despite the extensive work on visualisation in science education, there are at least two aspects of visualisation in science education that remain understudied. The first concerns the study of teachers' own visual representations, in particular how teachers' understanding of epistemic aspects of science can be captured in their visual representations. The second is related to the treatment of visual representations as "epistemic objects", not just cognitive and symbolic representations. Both of these aspects on teachers' visual representations are related to the epistemic dimensions of science. However, there hasn't been sufficient emphasis on visualisation of abstract notions such as NOS (Erduran, 2017), in particular how students might actually visualise the nature of "scientific knowledge" and "scientific practices" (Erduran & Kaya, 2018).

### 3.5.3   Analogies

The importance of analogies in science education has been highlighted for a few decades now (e.g. Duit, Roth, Komorek, & Wilbers, 2001; Glynn & Duit, 1995). An analogy is a comparison where a thing or a process is compared to another that is quite different from it with the aim of explaining the unfamiliar idea. Analogies and analogical reasoning underpin many scientific explanations. Historically, chemists used analogies to explain the phenomena that they were studying. For example, Boyle imagined elastic gas particles as moving coiled springs. According to Gentner (2002), "the basic intuition behind analogical reasoning is that when there are substantial parallels across different situations, there are likely to be further parallels" (p. 106). In this sense, analogical arguments can be used to generalise concepts, theories and methods so that they become applicable to classes of objects which are not of the same kind as those to which they originally apply (Tzanakis, 1998).

Gentner (1983) identifies three features of analogies: (a) literal similarity, a large number of both attributes and relations are mapped (e.g. the microphone is like our ear); (b) analogy, a large number of relations, but few attributes, are mapped (e.g. the hydrogen atom is like our solar system); and (c) abstraction, the base domain is an abstract relational structure (e.g. the hydrogen atom is a central force system). The notion that analogies usually cover target material that is difficult or abstract also supports the relative levels of abstraction of analog and target concepts. Analogies are generally used to make relational structure of the features of abstract target concepts more obvious to students than they would have been after a direct explanation of the target concept.

Science teachers, like scientists, frequently use analogies to explain concepts to students (James & Scharmann, 2007). Aubusson et al. (2009) argued that analogies help student learning by providing visualisation of an abstract concept, help to compare similarities of students' real world with new concepts and increase students' motivation. An influential framework in chemistry education was proposed by Johnstone who explained that learning and thinking in modern chemistry always take place in a constant shift between three different representational domains: the macroscopic, submicroscopic and symbolic domain. If these three domains (including the accompanying levels between the macroscopic and submicroscopic domains) and their interactions are misinterpreted, scientifically unreliable interpretations will necessarily emerge as a result (Johnstone, 1991). Analogies play a key role in the shifts between these domains. For example, balls might be used to represent the motion of molecules at evaporation.

## 3.6 Development of Pre-service Teachers' Epistemic Thinking

Strategies such as argumentation, visualisation and analogies can potentially support the learning of fairly abstract and difficult ideas and enhance the development of pre-service teachers' epistemic thinking. A further factor is how the subject knowledge itself is positioned in the learning of pre-service teachers. The epistemic core (i.e. aims and values, practices, methods and knowledge) is both a domain-general and a domain-specific framework. Each category of the epistemic core can be populated with specific topics from all science domains as well as concepts from a particular domain such as chemistry. In Chap. 2, we showed how the topic of acids and bases from chemistry can be used to relate particular practices and knowledge types such as models in chemistry. The same framework of the epistemic core can be applied to different topics, and through various iterations of its contextualisation, teachers' epistemic thinking can be consolidated. However, the question remains as to where to start in this process of integration of the epistemic core in teachers' learning.

*Should the epistemic core be taught first and then linked to science subject knowledge? Should the epistemic core and subject knowledge be taught concurrently? What is the optimal and most effective means of facilitating pre-service teachers learning? How do teachers' learning progress as they move across their career stages, from pre-service teacher education to in-service teaching practice?* These questions can be answered through empirical investigations by carrying out research on teachers' learning. They demand a research programme where alternative approaches can be empirically tested. An argument for a research agenda for incorporating epistemic ideas from philosophy of chemistry in teacher education was made in Chaps. 1 and 2. These chapters provided some content that teacher education can target as outcomes of pre-service and in-service teachers' learning. For instance, the specific nuances about aims and values of chemistry can be introduced to in-service teachers who have already appropriated understanding of epistemic aims and values of chemistry and why they are relevant for chemistry teaching. Nuanced outcomes may be too advanced for pre-service teachers who have yet to develop meta-perspectives on the subject that they are teaching. For the purposes of pre-service teacher education, it may be more fruitful to prioritise a domain-general understanding of the epistemic core first with some examples from chemistry which, in time, becomes more and more nuanced with more sophisticated chemistry content. It is indeed difficult to imagine how a chemically embellished epistemic core can be a starting point for pre-service teachers' learning before they don't understand what an epistemic core means in the first place or if they have limited chemistry knowledge. There are also questions related to how best to frame the learning of the epistemic core relative to teaching practice. Teacher education programmes vary in how they position theoretical knowledge about pedagogy and teaching practice in schools. The transformation of theoretical knowledge into teaching practice can be challenging for pre-service teachers (Cochran et al., 1993).

A further issue to consider with respect to the relationship between the epistemic core and subject knowledge of chemistry is that many pre-service teacher education programmes have expectations about pre-service teachers' background in chemistry. For instance, in England, applicants to the postgraduate certificate in education (PGCE) courses already have a bachelor's degree in chemistry, and they enrol in a 1-year course to develop pedagogical knowledge. Even in some systems like Turkey where empirical research was derived for the content of Chaps. 4, 5 and 6, chemistry courses are primarily clustered at the onset of teacher certification programmes with increasing emphasis on pedagogy in later years. Hence, in many educational systems from around the world, pre-service teachers will have had background in chemistry knowledge when they are following courses in a teacher education programme. Such arrangements in how subject knowledge and pedagogy are organised at an institutional level already present some constraints as to how the epistemic core can be related to pre-service teachers' learning of the subject knowledge. In other words, there is often already a separation of (a) subject knowledge from pedagogical knowledge and (b) epistemic thinking from subject knowledge and from pedagogical knowledge. Although some chemistry departments offer courses on philosophy of chemistry to chemistry undergraduates (e.g. see http://www.hyle.org/service/courses.htm for a list of international courses), these tend to be optional courses and not integrated by design into chemistry teaching and learning. Ideally, it would be beneficial for pre-service chemistry teachers to be introduced to the epistemic aspects of chemistry much earlier in their education, in undergraduate and postgraduate chemistry courses so that the subject teaching itself becomes more epistemically robust. In this sense, there is much work to be done in infusing the learning of meta-perspectives on chemistry not only in teacher education but also in undergraduate chemistry teaching as well suggesting implications not only for teacher educators but also for chemists. Although undergraduate chemistry teaching may be fairly similar in different national contexts, chemistry teacher preparation programmes can vary significantly depending on the national context.

## 3.7   Teacher Education in National Context

One challenge in recommending a model for infusing philosophy of chemistry in teacher education concerns the diversity of teacher education provision around the world, least of all due to the involvement of numerous institutions in the governance and audit of teacher education. Teacher education can also be centralised or decentralised depending on the national context. Australia and the USA are examples where teacher education is governed through a decentralised system. In Australia teacher registration is carried out by statutory teacher registration bodies in states and territories (Kleinhenz & Ingvarson, 2004). The registration bodies do not conduct any formal assessments in addition to, or separately from, those of the universities. On the other hand, in the USA which have a decentralised system of teacher education and certification, each state is responsible for initial credentialing of its teachers. Some states refer to this initial credential process as certification. Certification requirements vary greatly across the states, depending on local needs and available resources (Hirsch, Koppich, & Knapp, 2001).

The report entitled *Preparing Teachers Around the World* highlighted the diversity in teacher education provision from several countries in the world (Gitomer, 2003). It illustrated large differences in the scale of the teacher education enterprise across the countries included in the report. The surveyed countries were Australia, England, Hong Kong, Japan, Korea, Netherlands, Singapore and the USA. This report provides an exploratory analysis of teacher education and development policies in countries that participated in the *Repeat of the Third International Mathematics and Science Study at the Eighth Grade* (TIMSS, 1999) and scored as well as or higher than the USA. Among the centralised systems covered in Gitomer's report, teaching certification may be the responsibility of one government agency or multiple agencies. In Japan, Hong Kong, Korea and Singapore, the Ministry of Education or department of education governs almost all aspects of the teacher education and certification process. In the Netherlands, the responsibility is shared between the Ministry of Education and the Inspector of Education. The ministry sets minimum guidelines on entry and exit requirements and curriculum content for teacher education programs, while the inspector monitors compliance with these guidelines.

Most of the countries surveyed in Gitomer's (2003) report use high school grade point average and scores on national exit examinations taken in high school to select students for teacher education programmes. The structure and content of undergraduate teacher education programs were found to be quite similar across the countries surveyed. These include courses in subject area content, courses in educational theory and pedagogy and experiences observing and teaching students. Exit requirements are also similar across the countries surveyed. They typically include completion of an approved programme, tests and acceptable grades and student teaching experience. While all countries require student teaching experience as part of the teacher education curriculum, the duration of teaching practice ranges from 3 to 4 weeks in Japan to between 12 and 18 months in the Netherlands. The context of the teacher education programme where we implemented the intervention reported in Chaps. 4, 5 and 6 is similar to the countries surveyed by Gitomer (2003). It was also an undergraduate teacher education programme with similar course and exit requirements. The teaching practice experience expectation in the programme involves two courses, each lasting 14 weeks: (a) one based on observations of experienced teachers and (b) another based on own teaching of at least four lessons. Further details of this teacher education programme will be described later in the chapter.

In order to illustrate some of the more specific constraints involved in the incorporation of epistemic perspectives on chemistry in chemistry teacher education, two contrasting educational contexts are presented: the UK and Turkey. In the UK, there are broadly two approaches to teacher training: university-led teacher training and school-led teacher training. The university-led teacher training involves the 1-year postgraduate qualification, such as a PGCE (Postgraduate Certificate of Education) or PGDE (Postgraduate Diploma of Education, Scotland only) which are a common route into teaching across the UK. In Scotland, Northern Ireland and Republic of Ireland, a candidate must complete a postgraduate course in order to become a teacher. In the case of the school-led teacher training course, an applicant is typically selected by a school or group of schools. He or she is then in school from day 1, and most of the training will be based in that school. A lot of school-led courses also offer a postgraduate qualification. School Direct is a 1-year teacher training route available in England. *School-Centred Initial Teacher Training (SCITT)* is provided by groups of schools and colleges in England. In Wales, the *Graduate Teacher Programme (GTP)* gives people the opportunity to qualify as a teacher while working. *Teach First* is an employment-based route in England and Wales, aimed at recent graduates.

In Turkey, on the other hand, teacher education is carried out by university education faculties. Teacher candidates are trained based on a concurrent model in which they have both subject matter and teaching courses together. The courses include subject matter knowledge and skills in the proportion of 50–60%, knowledge and skills of the teaching profession in the proportion of 25–30% and general culture lessons with the proportion of 15–20%. Students commonly take teaching-related pedagogy courses including educational psychology, curriculum planning and teaching, measurement and evaluation and classroom management. Pre-service teachers mostly have teaching practice in the 4th year. These practices are carried out at cooperating schools under the supervision of cooperating teachers and instructors at faculties. However, the time dedicated to teaching practice has been changed over the years. Different from the elementary teacher education, most of the secondary school teaching including chemistry teaching would last for 5 years. Teacher candidates get their subject courses from the relevant faculties in their universities and pedagogical courses from the faculties of education. Students are admitted to these teacher education programs based on their scores from the national university entrance examinations. Only music, arts and physical education and sports teacher education programs apply additional ability tests while selecting their students. However, it should also be noted here that the students who graduate from the so-called Teacher High Schools (secondary schools that focus on education) get additional scores when they choose a teaching department. Anatolian Teacher High Schools are one of the high schools in Turkey, and they select their students through a national high school examination after elementary school (which is currently 12 years of compulsory education). There are now more than 200 Anatolian Teacher High Schools all around Turkey, and the main admission criterion to these high schools is the students' scores on a High School Entrance Examination. There are other routes to teacher education in Turkey. Another model involves obtaining a teaching certificate after the completion of a bachelor's degree in Science and Literature Faculties. Graduates apply to teaching certificate programs provided by most state or private universities. No matter from which program they graduate, all teacher candidates are required to get a certain score from the state exam called KPSS (Exam for the Selection of Civil Servants) to be recruited as a teacher.

### 3.7.1 Contrast of Teacher Education Programmes at Oxford and Bogazici

The contrast of national context is extended to the particular cases of our own universities where we work as teacher educators. The purpose of this section is to illustrate the nuances in the way that teacher education is structured which may offer opportunities as well as constraints to incorporating new content. The teacher education programmes at Oxford and Bogazici demonstrate sharp differences. For example, Oxford's PGCE Science course that includes teacher training in chemistry lasts for 9 months, whereas Bogazici's programme lasts for 4 years. Oxford's interns have typically at least a bachelor's degree in chemistry or related field before they get enrolled in the PGCE course, whereas Bogazici interns with master's degree combines chemistry training at the same time as pedagogical training. The University of Oxford, England, and Bogazici University, Turkey, are ranked highly in international university rankings. In the 2018 Times Higher Education World University rankings, the University of Oxford was ranked first, and Bogazici University was in the 401–500th range and is one of the only five Turkish universities included in the ranking. Both universities have a reputable history of science teacher education in their national contexts. Oxford started involvement in teacher education in 1885 while Bogazici introduced it in 1982.

Oxford PGCE Science (PGCE Oxford, 2018) course aims to produce high-quality teachers of the sciences across the 11–19 age range who will not only become competent teachers but will quickly become innovative leaders in their field. Pre-service teachers or "interns" will gain expertise in the different strategies for teaching science and will get insights into the way that pupils learn across the whole range of attainments, aptitudes and pupil differences. Interns learn how to turn their own subject knowledge into a form that can be understood by pupils. They are encouraged to think critically about the aims and practicalities of teaching science in schools. To attain these goals, interns work with each other, the University tutors and their mentors in schools, and they are encouraged to take responsibility for their own learning. The learning is structured through workshops, seminars, discussions, focussed assignments, school-based activities and sympathetic, expert, supervision and support. The programme runs from September to June. In the Autumn Term, interns spend time at the university and, in their first placement school, sometimes spend days in both the university and the school during the same week. During these joint weeks, initially 3 days are spent in the university and 2 days in school, and later in the first term, the schedule changes to 2 days in the university and 3 days in school. The spring term consists primarily of school experience, and for the summer term, interns move to a second school so that they have the opportunity to consolidate and extend their understanding and experience of learning and teaching in a new school environment. School teachers and university tutors contribute their particular expertise to the interns' learning and professional development. The interns therefore are presented with differing perspectives and points of view. The course is not seen as an apprenticeship scheme in which interns learn to teach like their mentors, nor is it a theory-into-practice scheme in which ideas taught in the university are put into practice in school. Instead, the emphasis is on developing critical learners, thinking about different perspectives and testing out ideas for themselves in practice. This process of reflection and experience underlies the whole course. As interns become increasingly competent as a teacher, they are encouraged to take responsibility for their own professional development and develop their own philosophy of learning and teaching.

Bogazici's Department of Mathematics and Science Education offers a master's degree without Thesis in Teaching Chemistry. Basic components of undergraduate programs consist of courses with a practical focus. Graduate programs focus on advanced theoretical knowledge in the field and on specialised applied research; a well-balanced emphasis on theory and application is maintained throughout the programs. Each program has a wide range of elective courses through which students receive a training program compatible with their interests and abilities. These programs, which have been developed to train students as competent teachers, educators and counsellors, also cater for students' physical, sensory and social needs. Therefore, students of the Faculty of Education are provided with the opportunity to attend sports, arts and literary events managed by various student clubs. Encouraging students to attend international exchange programs is also one of the basic principles of the faculty. While the chemistry teacher education programme lasts 9 months at Oxford, the duration of chemistry teacher education programme at Bogazici is 4 years. The teacher education intervention to be introduced in Chap. 4 was conducted when the programme used to last 5 years. Since 2014, the duration of chemistry teacher education in Turkey shifted from 5 years to 4 years. However the pre-service teachers participating in our project were part of the original cohort who continued to complete their programme in 5 years. In terms of university courses, chemistry teacher candidates at two universities take some common courses such as curriculum, assessment, instruction and lesson planning. On the other hand, the programmes include some different courses. For instance, chemistry teacher candidates at Bogazici complete chemistry courses such as general chemistry, physical chemistry, inorganic chemistry, organic chemistry, calculus, physics and educational psychology as an introductory pedagogy course throughout the programme during their teacher training, while those at the University of Oxford complete English as an Additional Language (EAL) or Special Educational Need (SEN) seminars. The practical application courses also differ in the two universities. While the candidates at Oxford enrol in practical science methods in application to teaching, chemistry, physics and biology (i.e. all chemistry teachers are trained to teach the three subjects to students up to age 16), this course for the candidates at Bogazici is only chemistry laboratory applications.

Pre-service teachers in both universities are placed into two different schools throughout the programme for their teaching practice. However, these universities have different regulations in terms of school-based mentoring. For example, chemistry teacher candidates at Oxford work with one science (or chemistry) mentor and one general education mentor, while those at Bogazici work with one chemistry teacher mentor in each of their placements. In other words, there is no designated general education mentorship based in schools associated with Bogazici. Oxford's component of this aspect of training includes expectations from interns to carry out investigations on various issues such as pastoral care, gender and equity as well as diversity and differentiation. The teacher candidates at Bogazici also enrol in a specific course with respect to their internships which they complete in their last year which consists of two semesters in the programme. They enrol in a course entitled *Practice Teaching in Chemistry* throughout their last semester which lasts 14 weeks. During this course, they discuss their classroom observations and reflections on what they did and observed in schools, with their peers in this course. Another difference between chemistry teacher education programmes at the two universities is related to History and Philosophy of Science (HPS) content taught to teacher candidates. Whereas there are options at Bogazici to take some HPS-related courses, there is none at Oxford. The PGCE Science course at Oxford has a very tight timeline of less than a year where the interns spend about two-thirds of their time in schools leaving a very limited schedule for university-based training which tend to cover more basic pedagogy-related content such as behaviour management, lesson planning and assessment.

## 3.8  Conclusions

The chapter contextualised teachers' learning and epistemic beliefs by illustrating the diversity in the governance of teacher education in different national contexts. It has been widely documented that teachers' epistemic beliefs can have an impact on students' learning (e.g. Sosu & Gray, 2012) and students' epistemic thinking can be supported by metacognitive prompts (e.g. Peters & Kitsantas, 2010). Although many research studies illustrated that student teachers' knowledge and epistemological views may be changed (Marra, 2005), there is limited understanding of how epistemology of chemistry itself plays out in pre-service teachers' minds. Teaching aspects of the epistemic nature of chemistry poses challenges to teachers, both pre- and in-service teachers. First, it requires the teaching of aspects of chemistry with which teachers have little familiarity. Second, teaching of nature of science in general often requires the use of pedagogic techniques such as orchestration of small group discussions that are still underused in many science classrooms. In addition, lesson materials that support teaching about the epistemic aspects of chemistry are scarce.

Furthermore, teachers may have misconceptions regarding aspects of nature of science that they are teaching. For example, in terms of models, teachers conceive scientific models in mechanical terms and believe that models are true pictures of non-observable phenomena and ideas (Gilbert, 1998). Research in science education has suggested a possible link between teachers' views of scientific knowledge and their classroom practice. Duschl and Wright (1989) found that the science teachers committed to a hypothetical-deductive view of the scientific method were also committed to teaching the propositional knowledge of the discipline. These teachers gave little consideration to the nature and role of theories in making curricular and instructional decisions. Smith and Anderson (1984) reported that a science teacher who believed that currently accepted scientific theories could be inferred from observation was surprised when her students failed to discover photosynthesis by observing the growth of plants. Lantz and Kass (1987) found that three high school chemistry teachers who used the same chemistry curriculum taught very different lessons about the nature of science, as a result of differences in their understanding of the nature of chemistry.

In order to educate future teachers about nature of chemistry in general and the epistemic core of chemistry in particular, teacher educators need to utilise strategies that would help unpack and develop epistemic thinking. This assumes that teacher educators themselves would be sufficiently familiar with the significance of the teaching and learning of epistemic themes in chemistry education. In Chap. 7, we will highlight some of our reflections and observations as teacher educators on how our own identities as teacher educators and routes into teacher education have influenced what we include in our teaching in teacher education programmes. As this chapter has illustrated, there are also systemic and institutional constraints surrounding teacher education programmes in universities which may hinder incorporation of novel instructional content. However, existing teacher education provision can be enriched through some strategies to begin the work of incorporating epistemic thinking in chemistry teachers' preparation.

The use of argumentation, visualisation and analogies, for example, can serve such a purpose in not only facilitating pre-service chemistry teachers' learning of epistemic aspects of chemistry but also can potentially instill in them the implied interactive pedagogical strategies that can eventually help with their students' learning if appropriated and applied in teaching. Future teachers need to learn to operate within the "language" of the scientific community (Mortimer & Scott, 2003) including the modes of reasoning and thinking captured through argumentation (Erduran & Jimenez-Aleixandre, 2007). They will need to be supported with explicit and useful tools including visual representations (e.g. Eilam & Gilbert, 2014) that have the potential to foster their metacognitive skills. The use of analogies can be considered a fruitful bridge between pre-service teachers' own ideas and the ideas about nature of chemistry that are targeted as learning outcomes in teacher education. Theoretical perspectives on pre-service teachers' epistemic beliefs and learning, along with some well-researched strategies like argumentation, can contribute to understanding of how best to design teacher education courses that support pre-service teachers' epistemic thinking in chemistry.

# References

Abd-El-Khalick, F., & Lederman, N. G. (2000). The influence of history of science courses on students' views on nature of science. *Journal of Research in Science Teaching, 37*(10), 1057–1059.

Akerson, V. L., Abd-El-Khalick, F., & Lederman, N. G. (2000). Influence of a reflective explicit activity-based approach on elementary teachers' conceptions of nature of science. *Journal of Research in Science Teaching, 37*(4), 295–317.

Akerson, V. L., Morrison, J. A., & Roth McDuffie, A. (2006). One course is not enough: Preservice elementary teachers' retention of improved views of nature of science. *Journal of Research in Science Teaching, 43*, 194–213.

Aubusson, P., Treagust, D., & Harrison, A. (2009). Learning and teaching science with analogies and metaphors. In *The world of science education: Handbook of research in Australasia*. Rotterdam, the Netherlands/Boston, MA: Sense Publishers.

Bay, J. M., Reys, B. J., & Reys, R. E. (1999). The top 10 elements that must be in place to implement standards-based mathematics curricula. *Kappan, 80*, 503–512.

Bell, R., Blair, L., Lederman, N. G., & Crawford, B. (2003). Just do it? Impact of a science apprenticeship on high school students' understandings of the nature of science and scientific inquiry. *Journal of Research in Science Teaching, 40*(5), 487–509.

Bondy, E., Ross, D., Adams, A., Nowak, R., Brownell, M., Hoppey, D., et al. (2007). Personal epistemologies and learning to teach, teacher education and special education. *The Journal of the Teacher Education Division of the Council for Exceptional Children, 30*(2), 67–82.

Brown, A. L, & Campione, J. C. (1990). Communities of learning and thinking, or a context by any other name. In D. Kuhn (Ed.), Developmental perspectives on teaching and learning thinking skills (special issue). *Contribution to Human Development, 21*, 108–126.

Buehl, M. M., & Alexander, P. A. (2002). Beliefs about schooled knowledge: Domain specific or domain general? *Contemporary Educational Psychology, 27*(3), 415–449.

Buehl, M. M., & Fives, H. (2016). The role of epistemic cognition in teacher learning and praxis. In J. A. Greene, W. A. Sandoval, & I. ten Bra (Eds.), *Handbook of epistemic cognition* (pp. 247–264). New York: Routledge.

Chambers, D. W. (1983). Stereotypic images of the scientist: The draw-a-scientist test. *Science Education, 6*, 255–265.

Chinn, C., Buckland, L., & Samarapungavan, A. (2011). Expanding dimensions of epistemic cognition: Arguments from philosophy and psychology. *Educational Psychologist, 46*, 141–167.

Chinn, C., & Rinehart, R. W. (2016). Epistemic cognition and philosophy: Developing a new framework for epistemic cognition. In J. A. Greene, W. A. Sandoval, & I. Braten (Eds.), *Handbook of epistemic cognition* (pp. 460–478). New York: Routledge.

Cochran, K. F., DeRuiter, J. A., & King, R. A. (1993). Pedagogical content knowing: An integrative model for teacher preparation. *Journal of Teacher Education, 44*, 263–272.

Dimopoulos, K., Koulaidis, V., & Sklaveniti, S. (2003). Towards an analysis of visual images in school science textbooks and press articles about science and technology. *Research in Science Education, 33*, 189–216.

Dori, Y. J., & Barak, M. (2001). Virtual and physical molecular modeling: Fostering model perception and spatial understanding. *Educational Technology & Society, 4*(1), 61–74.

Duit, R., Roth, W. M., Komorek, M., & Wilbers, J. (2001). Fostering conceptual change by analogies – between Scylla and Carybdis. *Learning and Instruction, 11*(4), 283–303.

Duschl, R. (1990). *Restructuring science education. The importance of theories and their development*. New York: Teachers College Press.

Duschl, R. (2008). Science education in 3-part harmony: Balancing conceptual, epistemic and social learning goals. *Review of Research in Education, 32*, 268–291.

Duschl, R. A., & Gitomer, D. H. (1997). Conceptual change in science and in the learning of science. In B. J. Fraser & K. G. Tobin (Eds.), *The international handbook of science education* (pp. 1047–1065). Dordrecht, The Netherlands: Kluwer Academic Publishers.

Duschl, R. A., & Wright, E. (1989). A case study of high school teachers' decision-making models for planning and teaching science. *Journal of Research in Science Teaching, 26*, 467–502.

Eilam, B., & Gilbert, J. K. (2014). *Science teachers' use of visual representations*. Dordrecht, The Netherlands: Springer.

Eilam, B., Poyas, Y., & Hashimshoni, R. (2014). Representing visually: What teachers know and what they prefer. In B. Eilam & J. K. Gilbert (Eds.), *Science teachers' use of visual representations* (pp. 53–83). Dordrecht, The Netherlands: Springer International Publishing.

Erduran, S. (2006). Promoting ideas, evidence and argument in initial teacher training. *School Science Review, 87*(321), 45–50.

Erduran, S. (2017). Visualising the nature of science: Beyond textual pieces to holistic images in science education. In K. Hahl, K. Juuti, J. Lampiselkä, J. Lavonen, & A. Uitto (Eds.), *Cognitive and affective aspects in science education research: Selected papers from the ESERA 2015 conference* (pp. 15–30). Dordrecht: Springer.

Erduran, S. (Ed.). (2019). *Argumentation in chemistry education: Research, policy and practice*. London: Royal Society of Chemistry.

Erduran, S., Aduriz-Bravo, A., & Mamlok-Naaman, R. (2007). Developing epistemologically empowered teachers: Examining the role of philosophy of chemistry in teacher education. *Science & Education, 16*(9–10), 975–989.

Erduran, S., & Dagher, Z. (2014). *Reconceptualizing the nature of science for science education: Scientific knowledge, practices and other family categories*. Dordrecht, The Netherlands: Springer.

Erduran, S., & Jimenez-Aleixandre, J. M. (2012). Research on argumentation in science education in Europe. In D. Jorde & J. Dillon (Eds.), *Science education research and practice in Europe: Retrospective and prospective* (pp. 253–289). Rotterdam, The Netherlands: Sense Publishers.

Erduran, S., & Jimenez-Aleixandre, M. P. (2007). *Argumentation in science education: Perspectives from classroom-based research*. Dordrecht, The Netherlands: Springer.

Erduran, S., & Kaya, E. (2018). Drawing nature of science in pre-service science teacher education: Epistemic insight through visual representations. *Research in Science Education, 48*(6), 1133–1149. https://doi.org/10.1007/s11165-018-9773-0

Erduran, S., Ozdem, Y., & Park, J. Y. (2015). Research trends on argumentation in science education: a journal content analysis from 1998–2014. *International Journal of STEM Education, 2015*(2), 5. https://doi.org/10.1186/s40594-015-0020-1

Evagorou, M., Erduran, S., & Mantyla, T. (2015). The role of visual representations in scientific practices: from conceptual understanding and knowledge generation to 'seeing' how science works. *International Journal of STEM Education, 2*, 11. https://doi.org/10.1186/s40594-015-0024-x

Flavell, J. H. (1979). Metacognition and cognitive monitoring: A new area of cognitive developmental inquiry. *American Psychologists, 34*, 906–911.

Friedrichsen, P., Van Driel, J. H., & Abell, S. K. (2010). Taking a closer look at science teaching orientations. *Science Education, 95*, 358–376.

Fullan, M. (2001). *The new meaning of educational change* (3rd ed.). New York: Teachers College Press.

Gentner, D. (1983). Structure-mapping: A theoretical framework for analogy. *Cognitive Science, 7*, 155–170.

Gentner, D. (2002). *Analogical reasoning, psychology of encyclopedia of cognitive science*. London: Nature Publishing Group.

Gess-Newsome, J. (2015). A model of teacher professional knowledge and skill including PCK: Results of the thinking from the PCK summit. In A. Berry, P. Friedrichsen, & J. Loughran (Eds.), *Re-examining pedagogical content knowledge in science education* (pp. 28–42). New York: Routledge.

Giere, R. N. (1999). *Science without laws*. Chicago: University of Chicago Press.

Gilbert, J. (1998). Explaining with models. In M. Ratcliffe (Ed.), *ASE guide to secondary science education*. London: Stanley Thornes.

Gilbert, J. K. (2005). Visualization: A metacognitive skill in science and science education. In J. K. Gilbert (Ed.), *Visualization in science education* (pp. 9–27). Dordrecht, The Netherlands: Springer.

Gilbert, J. K. (2010). The role of visual representations in the learning and teaching of science: An introduction. *Asia-Pacific Forum on Science Learning and Teaching, 11*(1), 1.

Gilbert, J. K., Reiner, M., & Nakhleh, M. (Eds.). (2008). *Visualisation: Theory and practice in science education*. New York/London: Springer.

Gitomer, D. (2003). *Preparing teachers around the world. Policy information report*. Princeton, NJ: Educational Testing Service.

Glynn, S. M., & Duit, R. (1995). Learning science meaningfully: Constructing conceptual models. In S. M. Glynn & R. Duit (Eds.), *Learning science in the schools: Research reforming practice* (pp. 3–33). Mahwah, NJ: Erlbaum.

Grandy, R., & Duschl, R. (2007). Reconsidering the character and role of inquiry in school science: Analysis of a conference. *Science & Education, 16*(1), 141–166.

Greene, J. A., Azevedo, R., & Torney-Purta, J. (2008). Modelling epistemic and ontological cognition: Philosophical perspectives and methodological directions. *Educational Psychologist, 43*, 142–160.

Greene, J. A., Sandoval, W. A., & Braten, I. (2016). *Handbook of epistemic cognition*. New York: Routledge.

Grossman, P. L. (1990). *The making of a teacher: Teacher knowledge and teacher education*. New York: Teachers College Press.

Habermas, J. (1981). *The theory of communicative action*. Boston: Beacon Press.

Haukoos, G. D., & Penick, J. E. (1985). The effects of classroom climate on college science students: A replication study. *Journal of Research in Science Teaching, 22*(2), 163–168.

Hirsch, E., Koppich, J. E., & Knapp, M. S. (2001). *Revisiting what states are doing to improve the quality of teaching: An update on patterns and trends*. Seattle, WA: University of Washington, Center for the Study of Teaching and Policy.

Hoban, G. F. (2002). *Teacher learning for educational change: A systems thinking approach*. Buckingham, UK: Open University Press.

Hofer, B. K. (2016). Epistemic cognition as a psychological construct: Advancements and challenges. In J. A. Greene, W. A. Sandoval, & I. Braten (Eds.), *Handbook of epistemic cognition* (pp. 19–38). New York: Routledge.

Hofer, B. K., & Pintrich, P. R. (1997). The development of epistemological theories: Beliefs about knowledge and knowing and their relation to learning. *Review of Educational Research, 67*(1), 88–140.

Ingersoll, R. (2003, September). *Is there really a teacher shortage?* (CPER Report #RR-03-4). Seattle, WA: A National Research Consortium, University of Washington.

Izquierdo, M., & Adúriz-Bravo, A. (2003). Epistemological foundations of school science. *Science & Education, 12*(1), 27–43.

James, M. C., & Scharmann, L. C. (2007). Using analogies to improve the teaching performance of preservice teachers. *Journal of Research in Science Teaching, 44*(4), 565–585.

Jimenez-Aleixandre, M. P., & Erduran, S. (2007). Argumentation in science education: An overview. In S. Erduran & M. P. Jimenez-Aleixandre (Eds.), *Argumentation in science education: Perspectives from classroom-based research* (pp. 3–27). Dordrecht, The Netherlands: Springer.

Johnstone, A. H. (1991). Why is science difficult to learn? Things are seldom what they seem. *Journal of Computer Assisted Learning, 7*(2), 75–83.

Joram, E. (2007). Clashing epistemologies: Aspiring teachers', practicing teachers', and professors' beliefs about knowledge and research in education. *Teaching and Teacher Education, 23*(2), 123–135.

Kang, N. (2008). Learning to teach science: Personal epistemology, teaching goals, and practices of teaching. *Teaching and Teacher Education, 24*, 478–498.

Kelly, G. J., Druker, S., & Chen, C. (1998). Students' reasoning about electricity: Combining performance assessments with argumentation analysis. *International Journal of Science Education, 20*(7), 849–871.

Kind, V. (2009). Pedagogical content knowledge in science education: Perspectives and potential for progress. *Studies in Science Education, 45*(2), 169–204.

Kitchener, R. F. (2002). Folk epistemology: An introduction. *New Ideas in Psychology, 20,* 89–105.

Kleinhenz, E., & Ingvarson, L. (2004). Teacher accountability in Australia: Current policies and practices and their relation to the improvement of teaching and learning. *Research Papers in Education, 19*(1), 31–49.

Kuhn, D., & Crowell, A. (2011). Dialogic argumentation as a vehicle for developing young adolescents' thinking. *Psychological Science, 22*(4), 545–552.

Lampert, M. (1990). When the problem is not the question and solution is not the answer: Mathematical knowing and teaching. *American Educational Research Journal, 27*(1), 29–64.

Lantz, O., & Kass, H. (1987). Chemistry teachers' functional paradigms. *Science Education, 71,* 117–134.

Leach, J., & Scott, P. (2002). Designing and evaluating science teaching sequence: An approach drawing upon the concept of learning demand and a social constructivist perspective on learning. *Studies in Science Education, 38,* 115–142.

Leach, J. T., Hind, A. J., & Ryder, J. (2003). Designing and evaluating short teaching interventions about the epistemology of science in high school classrooms. *Science Education, 87*(6), 831–848.

Lehrer, R., & Schauble, L. (2012). Supporting inquiry about the foundations of evolutionary thinking in the elementary grades. In S. M. Carver & J. Shrager (Eds.), *The journey from child to scientist: Integrating cognitive development and the education sciences* (pp. 171–206). Washington, DC: American Psychological Association.

Loucks-Horsley, S., Brooks, J. G., Carlson, M. O., Kuerbis, P. J., Marsh, D. D., & Padilla, M. J. (1990). *Developing and supporting teachers for science education in the middle years.* Andover, MA: National Center for Improving Science Education.

Loucks-Horsley, S., Hewson, P. W., Love, N., & Stiles, K. E. (1998). *Designing professional development for teachers of science and mathematics.* Thousand Oaks, CA: Corwin Press.

Loughran, J. (2007). Science teacher as learner. In N. G. Lederman & S. K. Abell (Eds.), *Handbook of research on science education* (pp. 1043–1065). New York: Routledge Taylor and Francis Group.

Luft, J. A., Firestone, J. B., Wong, S. S., Ortega, I., Adams, K., & Bang, E. (2011). Beginning secondary science teacher induction: A two-year mixed methods study. *Journal of Research in Science Teaching, 48*(10), 1199–1224.

Lunn Brownlee, J., & Schraw, G. (2016). Reflection and reflexivity: Higher order thinking in teachers' personal epistemologies. In G. Schraw, J. Brownlee, J. L. Olafson, & M. Vander Veldt (Eds.), *Teachers' personal epistemologies: Evolving models for transforming practice.* Charlotte, NC: Information Age Press.

Lunn Brownlee, J., Schraw, G., & Berthelsen, D. (2011). Personal epistemology and teacher education: An emerging field of research. In J. Brownlee, G. Schraw, & D. Berthelsen (Eds.), *Personal epistemology and teacher education* (pp. 3–21). New York: Routledge.

Magnusson, S., Krajcik, J., & Borko, H. (1999). Nature, sources and development of pedagogical content knowledge for science teaching. In J. Gess-Newsome & N. G. Lederman (Eds.), *Examining pedagogical content knowledge: The construct and its implication for science education* (pp. 95–132). Dordrecht, The Netherlands: Kluwer Academic.

Maloney, J., & Simon, S. (2006). Mapping children's discussions of evidence in science to assess collaboration and argumentation. *International Journal of Science Education, 28,* 1817–1841.

Marra, R. (2005). Teacher beliefs: The impact of the design of constructivist learning environments on instructor epistemologies. *Learning Environments Research, 8,* 135–155.

Mason, L. (1996). An analysis of children's construction of new knowledge through their use of reasoning and arguing in classroom discussions. *International Journal of Qualitative Studies in Education, 9*(4), 411–433.

Mayer, R. (2005). Multimedia learning: Guiding visuospatial thinking with instructional animation. In *The Cambridge handbook of visuospatial thinking* (pp. 477–508). Cambridge, MA: Cambridge University Press.

McComas, W. F. (2014). Analogies in science teaching. In W. F. McComas (Ed.), *The language of science education*. Rotterdam, The Netherlands: Sense Publishers.

Minstrell, J., & Van Zee, E. (Eds.). (2000). *Teaching in the inquiry-based science classroom*. Washington, DC: American Association for the Advancement of Science.

Mortimer, E. F., & Scott, P. H. (2003). *Meaning making in secondary science classrooms*. Maidenhead, UK: Open University Press.

Newton, P., Driver, R., & Osborne, J. (1999). The place of argumentation in the pedagogy of school science. *International Journal of Science Education, 21*(5), 553–576.

Ortwein, M., McCullough, A. C., & Thompson, A. (2015). A qualitative analysis of teachers' understandings of the epistemic aims of education. *Journal of Education and Human Development, 4*(3), 161–168.

Perry, W. G. (1970). *Forms of intellectual and ethical development in college years*. New York: Academic.

Peters, E. E., & Kitsantas, A. (2010). Self-regulation of student epistemic thinking in science: The role of metacognitive prompts. *Educational Psychology, 30*(1), 27–52. https://doi.org/10.1080/01443410903353294

Post, T. R., & Cramer, K. A. (1989). Knowledge, representation, and quantitative thinking. In M. C. Reynolds (Ed.), *Knowledge base for the beginning teacher* (pp. 221–232). New York: Pergamon.

Richmond, G., & Striley, J. (1996). Making meaning in classrooms: Social processes in small-group discourse and scientific knowledge building. *Journal of Research in Science Teaching, 33*, 839–858.

Roehrig, G. H., & Luft, J. A. (2004). Inquiry teaching in high school chemistry classrooms: The role of knowledge and beliefs. *Journal of Chemical Education, 81*(10), 1510–1516.

Rollnick, M., Bennett, J., Rhemtula, M., Dharsey, N., & Ndlovu, T. (2008). The place of subject matter knowledge in pedagogical content: A case study of South African teachers teaching the amount of substance and chemical equilibrium. *International Journal of Science Education, 30*(10), 1365–1387.

Schommer, M. (1990). Effects of beliefs about the nature of knowledge on comprehension. *Journal of Educational Psychology, 82*(3), 498.

Schraw, G., & Moshman, D. (1995). Metacognitive theories. *Educational Psychology Review, 7*(4), 351–371.

Schwab, J. (1962). The teaching of science as enquiry. In J. J. Schwab & P. F. Brandwein (Eds.), *The teaching of science* (pp. 1–103). New York: Simon and Schuster.

Schwartz, R. S., & Lederman, N. G. (2002). "It's the nature of the Beast": The influence of knowledge and intentions on learning and teaching nature of science. *Journal of Research in Science Teaching, 39*(3), 205–236.

Schwartz, R. S., Lederman, N. G., & Crawford, B. A. (2004). Developing views of nature of science in an authentic context: An explicit approach to bridging the gap between nature of science and scientific inquiry. *Science Education, 88*(4), 610–645.

Scott, P., Leach, J., Hind, A., & Lewis, J. (2006). Designing research evidence-informed teaching strategy. In R. Millar, J. Leach, J. Osborne, & M. Ratcliffe (Eds.), *Improving subject teaching: Lessons from research in science education* (pp. 60–78). London: Routledge.

Sendur, G., Polat, M., & Kazanci, C. (2017). Does a course on the history and philosophy of chemistry have any effect on prospective chemistry teachers' perceptions? The case of chemistry and the chemist. *Chemistry Education Research and Practice, 18*, 601–629.

Shulman, L. S. (1986). Those who understand: Knowledge growth in teaching. *Educational Researcher, 15*(2), 4–14.

Shulman, L. S. (1987). Knowledge and teaching: Foundations of the new reform. *Harvard Educational Review, 57*(1), 1–22.

Simon, S., Erduran, S., & Osborne, J. (2006). Learning to teach argumentation: Research and development in the science classroom. *International Journal of Science Education, 28*(2), 235–260.

Simon, S., Osborne, J., & Erduran, S. (2003). Systemic teacher development to enhance the use of argumentation in school science activities. In J. Wallace & J. Loughran (Eds.), *Leadership and professional development in science education: New possibilities for enhancing teacher learning* (pp. 198–217). London/New York: RoutledgeFalmer.

Smith, E. L., & Anderson, C. W. (1984). Plants as producers: A case study of elementary science teaching. *Journal of Research in Science Teaching, 21*(7), 685–698.

Smylie, M. A. (1989). Teachers' view of the effectiveness of sources of learning to teach. *Elementary School Journal, 89*(5), 543–558.

Sosu, E. M., & Gray, D. S. (2012). Investigating change in epistemic beliefs: An evaluation of the impact of student teachers' beliefs on instructional preference and teaching competence. *International Journal of Educational Research, 53*, 80–92.

Spillane, J. S. (1999). External reform initiatives and teachers' efforts to reconstruct their practice: The mediating role of teachers' zones of enactment. *Journal of Curriculum Studies, 31*(2), 143–175.

Supovitz, J. A., & Turner, H. M. (2000). The effects of professional development on science teaching practices and classroom culture. *Journal of Research in Science Teaching, 37*(9), 963–980.

TIMSS. (1999). *International science report: Findings from IEA's repeat of the third international mathematics and science study at the eighth grade.* Retrieved from https://timss.bc.edu/timss1999i/math_achievement_report.html

Toulmin, S. (1958). *The uses of argument.* Cambridge, UK: Cambridge University Press.

Tsai, C. C. (2007). Teachers' scientific epistemological views: The coherence with instruction and students' views. *Science Education, 91*(2), 222–243.

Tzanakis, C. (1998). Discovering by analogy: The case of Schrödinger's equation. *European Journal of Physics, 19*, 69–75.

Veal, W. R. (2004). Beliefs and knowledge in chemistry teacher development. *International Journal of Science Education, 26*(3), 329–351.

Vesterinen, V. M. (2012). *Nature of science for chemistry education: design of chemistry teacher education course.* Unpublished PhD thesis. Helsinki: University of Helsinki.

Yadav, A., Herron, M., & Samarapungavan, A. (2011). Personal epistemology in preservice teacher education. In J. Lunn Brownlee, G. Schraw, & D. Berthelsen (Eds.), *Personal epistemology and teacher education* (pp. 25–39). New York: Routledge.

Zembal-Saul, C. (2009). Learning to teach elementary school science as argument. *Science Education, 93*(4), 687–719.

Zembal-Saul, C., & Vaishampayan, A. (2019). Research and practice on science teachers' continuous professional development in argumentation. In S. Erduran (Ed.), *Argumentation in chemistry education: Research, policy and practice* (pp. 142–172). London: Royal Society of Chemistry.

Zembal-Saul, C., Munford, D., Crawford, B., Friedrichsen, P., & Land, S. (2002). Scaffolding preservice science teachers' evidence-based arguments during an investigation of natural selection. *Research in Science Education, 32*, 437–463.

Zohar, A. (2008). Science teacher education and professional development in argumentation. In S. Erduran & M. P. Jimenez-Aleixandre (Eds.), *Argumentation in science education: Perspectives from classroom-based research* (pp. 245–268). Dordrecht, The Netherlands: Springer.

Zohar, A. (2012). Explicit teaching of metastrategic knowledge: Definitions, students' learning, and teachers' professional development. In A. Zohar & Y. J. Dori (Eds.), *Metacognition in science education: Trends in current research* (pp. 197–223). Dordrecht, The Netherlands: Springer.

Zohar, A., & Ben-David, A. (2008). Explicit teaching of meta-strategic knowledge in authentic classroom situations. *Metacognition Learning, 3*, 59–82.

# Chapter 4
# Incorporating the Epistemic Core in Teacher Education Practice

## 4.1 Introduction

A vision for incorporating novel learning goals such as the learning of epistemic aspects of chemistry demands significant shifts in what is expected of teaching and teachers. In contemporary educational landscape, an example novel vision for teaching and learning of science has been outlined in the recent *Next Generation Science Standards* (NGSS) (NGSS Lead States, 2013) from the USA. The design of NGSS has been informed by science education and cognitive science research about what it means to teach and learn science effectively. NGSS has three interrelated goals: (a) core ideas, pointing to the shift in the emphasis away from the breadth of too much content to a focus on the in-depth development of core explanatory ideas; (b) practices through which students develop key explanatory ideas and models through investigation and apply them to make sense of phenomena; and (c) coherence or building explanatory ideas requires treating science learning as a coherent progression in which learners build ideas across time and between science disciplines. Our approach to the teaching and learning of the epistemic core of chemistry is consistent with the pedagogical vision that is implicit in and is promoted by the NGSS. Our position is that the core ideas of chemistry need to be unpacked in a way that exposes the disciplinary structures and processes to learners. The visual tools that are reviewed in Chap. 2 capture some key ideas on the nature of scientific knowledge (i.e. main categories of knowledge), scientific practices such as modelling, argumentation and representation (e.g. science and engineering practices) and interrelations within and across the epistemic core categories of aims and values, practices, methods and knowledge (i.e. coherence). As such, investigations into pre-service teachers' engagement with such epistemic aspects of chemistry can contribute to much needed research on early career teachers of science (e.g. Luft, 2007).

Reiser (2013) summarises some of the shifts that are required of teachers' knowledge and practice for adoption of NGSS recommendations. For example, a large

© Springer Nature Switzerland AG 2019
S. Erduran, E. Kaya, *Transforming Teacher Education Through the Epistemic Core of Chemistry*, Science: Philosophy, History and Education,
https://doi.org/10.1007/978-3-030-15326-7_4

part of the teachers' role is to support the knowledge building aspects of practices, not just the procedural skills in doing experiments, and hence extensive class focus needs to be devoted to argumentation and reaching consensus about ideas, rather than having textbooks and teachers present ideas to students. Although new tools including new curriculum materials such as NGSS and new assessments are important supports for the secondary education systems in these directions, without a strong focus on aligned professional development, adopting NGSS and providing these resources to teachers are not sufficient (Reiser, 2013).

How, then can teacher education be structured to facilitate pre-service teachers' understanding of novel curriculum content? There is considerable research on design principles and strategies for effective teacher education. Some of these principles and strategies were reviewed in Chap. 3. For example, research has focused on various aspects such as pre-service teachers' conceptions of teaching and learning science (Brickhouse & Bodner, 1992) and their subject-matter knowledge (Gess-Newsome, 1999). Supovitz and Turner (2000, p. 964) identified that high-quality teacher education must (a) immerse participants in inquiry, questioning and experimentation; (b) be intensive and sustained; (c) engage teachers in concrete teaching tasks and be based on teachers' experiences with students; (d) focus on subject-matter knowledge and deepen teachers' content skills; (e) be grounded in a common set of professional development standards and show teachers how to connect their work to specific standards; and (f) be connected to other aspects of school change. Apart from an understanding of the content (or subject) domain such as chemistry and the epistemology of the domain, teachers need to understand how to transform these notions into teachable scenarios (Loucks-Horsley, Hewson, Love, & Stiles, 1998). Since pre-service teachers are not yet integrated into the school culture full time, the likelihood of their involvement in participation in other aspects of school change is fairly low.

In this chapter, we present a teacher education module that was designed to incorporate the "epistemic core" idea into pre-service teachers' learning. In relation to the module, we were mindful of recommendations from the research literature about how best to maximise pre-service teachers' learning. However, we were also pragmatic about teaching fairly abstract and dense concepts within the very real constraints of teacher education in the particular context. For instance, within the timeframe of the project, the pre-service teachers were not due to do teaching practice in schools. Their existing time allowance in their internship schools was already taken up by programme-level requirements. Hence, engaging them in school-based tasks was not possible. It is likely that many initial teacher education programmes where modules and courses are added onto the existing programme of work will have similar constraints. We were also guided by the broader research problem of testing a "proof of concept". In other words, we were interested in finding out whether or not it is possible at all to infuse fairly abstract epistemic ideas in teacher education in a coordinated manner and what the outcome of such infusion would be, regardless of the extent to which they can be fully integrated in the teacher preparation programmes. In this respect, while the project is a realistic representation of what is possible to achieve in an initial teacher education programme in terms of reform efforts, it is also limited in terms of its full potential implementation. Overall the sort of work that is being reported in this book is part of a broader research programme that demand fairly detailed investigations to various aspects of teacher education and how teacher education can be optimised for the uptake of epistemic thinking by pre-service chemistry teachers. Not one project is likely to answer the plethora of research questions that are raised, including the impact of teacher education on teaching practice.

## 4.2   Teacher Education Context in Turkey

The module was part of a research and development project focusing on the nature of science (Erduran & Kaya, 2018; Kaya & Erduran, 2016; Kaya, Erduran, Aksoz, & Akgun, 2019). The project took place in a university pre-service teacher education programme and was offered as part of an undergraduate module at a state university in Turkey. In order to contextualise the project, some information on research and teacher education in Turkey will be presented. Chemistry education research in Turkey fares well internationally in terms of frequency of publications suggesting the presence of an active research community. According to Teo, Goh and Yeo (2014), Turkey was the second in the list of countries where the most number of research studies was conducted in chemistry education. These authors conducted content analysis of 650 empirical chemistry education research papers published in two top-tiered chemistry education journals *Chemistry Education Research and Practice* and *Journal of Chemical Education* and three top-tiered science education journals *International Journal of Science Education, Journal of Research in Science Teaching* and *Science Education* from 2004 to 2013. They found that the 60 empirical chemistry education research papers emerging from Turkey accounted for 9.2% of the papers, and it ranked second after 47.4% having been conducted in the USA. The proportion of research in chemistry teacher education, however, is relatively low overall in the literature. Teo and colleagues noted that research on pre-service chemistry teacher education accounted for 8.3% of papers. In a related study conducted directly on papers from Turkish authors, Erduran and Mugaloglu (2016) reported that only 9% of papers focused on history and philosophy of science and 3% on teacher education in an analysis of the top journals of *Journal of Research in Science Teaching, Science Education* and *International Journal of Science Education* from 1998 to 2012. Overall, then, although there is a healthy research culture in chemistry education in Turkey, the precise focus of history and philosophy of chemistry is fairly limited (e.g. Tumay, 2016).

   Research on recent policies about teacher education in Turkey indicate that since 1973, teachers have been educated in higher education institutions (Cakiroglu & Cakiroglu, 2003; YÖK, 2007). In 1989, the length of teacher education for all teacher education institutions, including 2-year education institutes training elementary level teachers, increased to at least 4 years with the decision of the Higher Education Council. As part of a doctoral thesis, Eret (2013) carried out a review of the teacher education provision in Turkey. She reported that there are a total of 72 education faculties under the Higher Education Council (HEC) in Turkey. Of these faculties, 64 of them are in state universities. Preschool and elementary school teacher education lasts for 4 years in faculties of education. The concurrent model of teacher education is used in which candidates take subject-matter and teaching courses together. The courses include subject-matter knowledge and skills (about 50–60%), knowledge and skills of the teaching profession (about 25–30%) and general culture lessons (about 15–20%). Eret further indicates that pre-service teachers commonly take teaching-related pedagogy courses, which are Methods of Teaching 1–2, Introduction to Educational Science, Educational Psychology, Curriculum Planning and Teaching, Measurement and Evaluation, Turkish Education System and School Management, Classroom Management, Guidance, Instructional Technologies and Materials Design, School Experience and Teaching Practice.

Teaching practice is typically dedicated to the 4th year of the teacher education programmes. These practices are carried out at cooperating schools under the supervision of mentor teachers and instructors at faculties. Different from the elementary teacher education, in previous years most of the secondary school teaching (Secondary Science and Mathematics, and Social Areas Teaching) lasts for 5 years but has now been reduced to 4 years in chemistry teaching (YÖK, 2018). Pre-service teachers get their subject courses from the relevant faculties in their universities and teaching courses from the faculties of education. In most universities in Turkey, pre-service teachers follow all their chemistry subject courses along with science pedagogy courses in faculties of education. Bogazici University – where the project reported in this chapter took place – has a different arrangement in that the teacher candidates take chemistry classes from the department of chemistry. (It should also be noted that the sample in our project were still subjected to the 5-year programme since the project took place prior to the new 4 year regulation coming into effect.) Chemistry teacher training programmes consist of ten semesters, and Special Teaching Methods-I and School Experience-I courses are taken in the eighth semester, and Special Teaching Methods-II and School Experience-II courses are taken in the ninth semester and Teaching Experience course is taken in the tenth semester. In the School Experience-I and School Experience-II courses, pre-service teachers only make observations about the teaching and learning process such as teaching methods (Akkus & Uner, 2017). While in some universities related courses such as nature of science in science education may be a compulsory course in science teacher education programmes, it is not required at Bogazici University. The teacher education module described in subsequent sections as well as in Chaps. 5 and 6 was offered as an elective course.

A total of 15 female pre-service teachers participated in the project. However, we will focus on data from one group of four pre-service teachers in Chaps. 5 and 6 to provide in-depth coverage of their learning. These pre-service teachers were senior year students. The teacher education intervention lasted for 11 3 h sessions. The sessions will be described in more detail in the rest of this chapter. Erduran and Dagher's (2014) book was used as a key resource in the module. As an example, during the session on aims and values, pre-service teachers were given a set of aims and values derived from Erduran and Dagher's (2014) work and were asked to draw posters to represent how they could link them to chemistry lessons. The details of this session will be described in the rest of the chapter.

In the session focusing on scientific practices, the participants were introduced to the Benzene Ring Heuristic (BRH) with an example of acid-base chemistry. Each aspect of the heuristic was reviewed, and pre-service teachers were asked to produce other examples in their groups to visually represent how they understand scientific practices. All sessions utilised a set of visual images that are collectively referred to as "Generative Images of Science" by Erduran and Dagher (2014) (see Chap. 2). These visual tools can potentially support meta-level understanding of nature of science (NOS). The importance of metacognitive training on pre-service teachers' understanding of NOS has been stressed in the literature (e.g. Duschl & Erduran, 1996). Pre-service teachers had a copy of each of these images throughout the sessions after, and they were encouraged to refer to them in the context of the tasks that they were engaged in for that particular session. Each session focusing on an epistemic core idea (where the instructors provided input) culminated in the production of some lesson resources in the subsequent session where pre-service teachers worked in groups. In these latter sessions, pre-service teachers reviewed what they had produced the previous week and used these resources along with others like common science textbooks and the Internet to produce a lesson plan. All sessions also used numerous active learning strategies such as group discussions, presentations and posters. In the rest of this chapter, we will detail the content of each session and illustrate the tasks and strategies that we developed and implemented.

## 4.3  Design of Teacher Education Sessions

The overall objective of the teacher education intervention was to facilitate pre-service teachers' learning to teach nature of chemistry. In particular, the sessions were designed to integrate the epistemic core idea introduced in Chap. 2 as part of nature of chemistry. The pre-service teachers were immersed in contexts where they could (a) critique different perspectives on the nature of science in science education; (b) understand the Family Resemblance Approach (FRA) to NOS including the definitions of the aims and values of science, methods and methodological rules, scientific practices and scientific knowledge as characterised in Erduran and Dagher's (2014) book; and (c) use the epistemic core to plan lessons and develop student resources. Although Erduran and Dagher's theoretical framework guided the design of the sessions, there was transformation of these authors' framework for the practical purposes of teacher education. This aspect of the work required not only the theoretical insight from Erduran and Dagher's work but also the research evidence from teacher education to inform the content of the sessions such that they are supportive of pre-service teachers' learning. Our knowledge of effective design principles of teacher education (e.g. Supovitz & Turner, 2000) and teachers' knowledge (e.g. Shulman, 1986) complemented the consideration of the epistemic core ideas in each session. For example, we were mindful of empowering the pre-service teachers in ways that would take ownership of the ideas, by encouraging them to apply the ideas that they learnt by producing resources. The pre-service teachers were free to draw on their own examples to illustrate issues discussed in the sessions. We tried to create a safe environment where the pre-service teachers could express their views and opinions without fearing that they would be judged for being wrong.

The sessions included plenty of opportunities for group discussions as well as production of artefacts to externalise and support the pre-service teachers' thinking. In each session focusing specifically on the epistemic core, particular themes were emphasised. For instance, in the scientific methods session, there was emphasis on diversity of methods. In the scientific knowledge session, there was focus on coherence between theories, laws and models. Finally, we were mindful of creating a "community of practice" (Lave & Wenger, 1991). Lave and Wenger formulated the notion of "communities of practice" among adults, studying how people in organisations share information and collaborate and learn from one another (Lave & Wenger, 1991; Wenger, 1998). Learners are coached through "cognitive apprenticeships", and learning is supported by knowledgeable others who can model, mentor and guide their understanding where learning is situated in a relevant and motivating context (Brown, Collins, & Duguid, 1989; Lave & Wenger, 1991; Wenger, McDermott, & Snyder, 2002). In designing and implementing teacher education programmes to infuse the epistemic core of chemistry, it will be vital to consider the theoretical arguments as well as the empirical evidence on what supports beginning teachers' learning.

An assumption in the design of the teacher education sessions was that the epistemic core of science is a meta-level characterisation of science, and as such its teaching and learning, be it for teachers or students, requires metacognition (e.g. Zohar, 2012). Zohar and Ben-David (2008) discuss meta-strategic knowledge or MSK as a subcomponent of metacognition. MSK is the "thinking behind the thinking" (meta-level of thinking) rather than the "thinking behind the doing" (Zohar & Ben-David, 2008). Explicit teaching of MSK refers not only to the thinking about what we do (procedural level of thinking) in the scientific inquiry processes but also to explicit thinking about how we think (i.e. a metacognitive level of thinking) in order to "do" the inquiry processes. Ben-David and Zohar (2009) argue that in order to carry out thoughtful inquiry processes, students need to explicitly understand how scientists think and why they think in that way, rather than only to know what scientists do and think during the scientific inquiry processes. Therefore "the concept MSK points to explicit meta-procedural understanding (rather than to procedural understanding) of scientific thinking strategies" (Ben-David & Zohar, 2009, p. 1660).

Consideration of MSK in relation to scientific inquiry by Zohar and Ben-David (2008) and Ben David and Zohar (2009) follows earlier work by Kuhn who defined meta-level knowledge as procedural and operating in the real selection and regulation of the use of inquiry strategies. Kuhn had emphasised the procedural nature of the meta-level by using the term "meta-strategic knowing" instead of "knowledge"

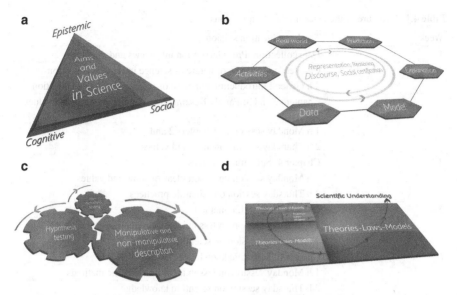

**Fig. 4.1** Generative Images of Science. (From Erduran & Dagher, 2014, p. 164)

in most of her work (Kuhn, 1999, 2001, 2002). Our incorporation of the *Generative Images of Science* (Erduran & Dagher, 2014) in the teacher education sessions intended to facilitate the development of pre-service teachers' MSK about the epistemic core of science through visual representations (see Fig. 4.1). Given the abstract nature of the concepts related to the epistemic aspects of chemistry, we anticipated that the use of visual tools can serve as cognitive tools in communicating epistemic content and processes and thus support pre-service teachers' epistemic thinking.

In the rest of this section, the content of each session will be described, and examples of the tasks used with pre-service teachers will be outlined. The outcomes of the particular emphasis on each theme of epistemic core sessions are illustrated in more depth in Chaps. 5 and 6. There were 11 teacher education sessions (see Table 4.1). There was an overlap of the reflection on one category and the introduction of another category, for example, reflecting and thinking about lesson ideas about Aims and Values on Monday and workshop-style activities on Scientific Practices on the Thursday of the same week (see Table 4.1). All teacher education courses at the university are covered over a total of a 14-week period. It is also typical that each session would be split into two sessions for a particular week. Hence, the particular decisions regarding the placement of the sessions were made on the basis of the overall programme requirements at the university. However, we were mindful of splitting the sessions within a particular week in order to provide periods of pedagogical reflection before introduction of another theme. Hence, we accommodated the goals of reflection and the programme requirements within the design of the teacher education sessions. Finally, the pre-service teachers were encouraged

**Table 4.1** Structure of the teacher education sessions

| Week | Teacher education session |
| --- | --- |
| 1 | Data collection: Pre-intervention interviews and drawings |
| 2 | Chapter 1: Overview of nature of science in science education |
| | 2 h General introduction to nature of science in science education |
| 3 | Chapters 2 and 3: Family Resemblance Approach; Aims and values of science |
| | 1 h Monday session on Chapters 2 and 3 |
| | 2 h Thursday session on aims and values |
| 4 | Chapter 4: Scientific practices |
| | 1 h Monday session on lesson ideas on aims and values |
| | 2 h Thursday session on scientific practices |
| 5 | Chapter 5: Scientific methods |
| | 1 h Monday session on lesson ideas on scientific practices |
| | 2 h Thursday session on scientific methods |
| 6 | Chapter 6: Scientific knowledge |
| | 1 h Monday session on lesson ideas on scientific methods |
| | 2 h Thursday session on scientific knowledge |
| 7 | Chapter 7: Social contexts |
| | 1 h Monday session on lesson ideas on scientific knowledge |
| | 2 h Thursday session on social contexts |
| 8 | Chapter 8: Generative Images of Science |
| | 1 h Monday session on lesson ideas on social contexts |
| | 2 h Thursday session on Generative Images of Science |
| 9–10 | Projects: Pre-service teachers prepare projects to collate all categories into lesson ideas |
| 11 | University holidays |
| 12–13 | Presentations of projects: Pre-service teachers present their projects |
| 14 | Data collection: Post-intervention interviews and drawings |

to bring together all categories in Weeks 9 and 10 to produce projects (i.e. lesson materials for school students) which they presented in class during Weeks 12 and 13. The post-data collection occurred in Week 14. The "epistemic core" sessions were covered in Weeks 3–7. Hence, there were 7 weeks from the end of the last session on this theme and the conduct of post-intervention data collection.

### 4.3.1   Session on Introduction to Nature of Science

The first session of the module included plenty of open-ended discussion to give the participants an opportunity to reflect on what they think about nature of science and to hear from their peers' ideas which might have been different from their own. The purpose of such an open-ended discussion was to establish where the pre-service teachers' were coming from and, as such, to inform the teacher educator about how best to mediate the discussion in order to accomplish the goals of the session. The session was guided by the following two key questions:

- *What comes to your mind when you hear "nature of science"?*
- *What ideas about nature of science should we teach in science lessons?*

Among all the sessions in the project, the first introductory session was the most theoretical in nature and was primarily based on discussions about NOS, its history and its presence in science curriculum, teaching and learning. Examples were drawn from all sciences to reflect about what makes the school sciences (i.e. biology, physics, chemistry) "science". Following on from an initial discussion in groups and then in the whole class, the pre-service teachers were asked to reflect on the history of NOS in science education. Chapter 1 from Erduran and Dagher's (2014) book had been a required reading for this session. A set of questions were displayed on the board to get them to think about the key researchers in science education having advanced arguments for the inclusion of NOS in science curriculum from the 1960s, followed by some more contemporary characterisations of NOS in science education. The participants would have read pages 3–9 from Chapter 1 of Erduran and Dagher (2014) that includes a focused discussion about the history of NOS. The pre-service teachers were encouraged to compare and contrast the different perspectives on NOS (e.g. Allchin, 2011; Lederman, Abd-El-Khalick, Bell, & Schwartz, 2002; Matthews, 2012) as outlined in Erduran and Dagher's book and discuss the pros and cons of each perspective. The intention with this discussion was (a) to lay the foundation of pre-service teachers' understanding of the theoretical background on NOS and (b) to encourage argumentation as part of a broader goal of education which is equipping future teachers with critical reasoning skills. As mentioned in Chap. 3, argumentation was one of the strategies used in the sessions.

### 4.3.2   Session on the Family Resemblance Approach

The participants were expected to read Chapter 2 from Erduran and Dagher's (2014) book which outlines the perspective of the Family Resemblance Approach (FRA) to NOS. Hence, the chapter is intended to set the theoretical background to the session. The participants were encouraged to produce reflections of their reading to bring to the session. The primary aim of the session was to present FRA and explore its implications for science teaching and learning. The main reason for including this session was to get the pre-service teachers to think about *how we know about how science works and why we categorise different sciences as "science"*. Before applying the idea of FRA, it was crucial for pre-service teachers to understand (a) what is meant by FRA and (b) how FRA is used as a means to justify why and how NOS is characterised, i.e. on the basis of a family group sharing particular features such as aims, values, method, practices and knowledge. In other words, chemistry, physics and biology (as well as other science disciplines) share certain characteristics (i.e. domain-general), and at the same time, they also have specific disciplinary focus (i.e. domain-specific). For an extended discussion of this issue, see Erduran and Dagher (2014).

The session began with a set of photographs of different families as represented in Fig. 4.2. The pre-service teachers were encouraged to consider the photographs and think about how the photographs represent different families. They were asked to discuss the following questions in groups of 4–5:

- *These are photographs of families. Can you think of what makes a "family"?*
- *What characteristics do members of a family have?*
- *What is a biological family? Are these biological families? Why?*
- *Are there other kinds of families?*

The aim of this activity is to focus on the concept of "family resemblance" by using biological family and particular characteristics such as eye and hair colour and facial characteristics that are shared by the members in a family. Other types of families, for instance, those based on non-biological partnerships and friendships, are also raised in order to clarify the criteria for selection of biological families. As a whole, the task is then used to build up an everyday analogy of how different sciences such as biology, chemistry and physics are part of a "family" because they resemble each other when considered relative to a set of criteria such as aims, values, methods and practices. Hence, the participants are asked to think of different branches of science and how/why they are grouped together, for instance, because they share certain characteristics such as aims and values, methods and methodological rules, practices, knowledge forms and social-institutional contexts. The overall objective of this session was to provide the pre-service teachers with oppor-

**Fig. 4.2** Examples of family photographs used in the Family Resemblance Approach session

tunities to explore how and why we conceptualise a particular science as "science" and what makes science "science". In the discussions, they would have been expected to think about common features of different branches of science (e.g. the importance of having accurate data) as well as some of differences across sciences (e.g. astronomy using historical data versus chemistry using data manipulated in the laboratory). In summary, the session was intended to set the scene for broad epistemological considerations of chemistry before zooming in on particular aspects such as aims and values, practices, methods and knowledge.

### 4.3.3 Session on Aims and Values of Science

The session on aims and values of science in general and in chemistry in particular is explored through a set of questions to promote discussion among pre-service teachers. The background reading was Chapter 3 on aims and values from Erduran and Dagher's (2014) book. The questions are intended to get pre-service teachers to think about what chemistry aims to accomplish and what epistemic values chemists might have. These may include social as well as cognitive and epistemic aims and values (see Fig. 4.1a). Further questions encourage pre-service teachers to link their ideas about the previous questions to educational settings by considering the implications of aims and values of chemistry for curriculum and instruction. Example questions include:

- *What are the aims and values of science?*
- *What is the significance of scientific aims and values in constructing scientific knowledge?*
- *Do you think scientific aims and values are part of the science curriculum? Can you think of an example that you might have seen in the curriculum?*
- *Do you think scientific aims and values can be promoted in science lessons? If yes, how?*

An example of an aim and a value in science including chemistry is objectivity, as reviewed in Chap. 2. In the context of chemistry, particular aims such as "objectivity" can be closely linked to economic ends in industry, as noted by Baird (2000). Hence, while the aim of objectivity might be universal in different sciences, its enactment in particular disciplines such as chemistry needs closer articulation in order to understand how objectivity works in chemistry. In other words, while some criteria such as "objectivity" might assign science family membership to chemistry, it may also illustrate what could potentially make its meaning particular in chemistry. Since our goal in this session was to instill in pre-service teachers the idea of epistemic aims and values, our emphasis was on the learning of what "epistemic" means and how it applies to chemistry. Nuance about what makes particular epistemic aims and values such as "objectivity" chemical can be discussed through the application of the family resemblance idea to chemistry and how to differentiate objectivity in chemistry from objectivity in other sciences. These are more advanced and sophisticated ideas and would follow from a more basic understanding of epistemic aims and values.

### 4.3.4   Session on Scientific Methods

The session on scientific methods required the reading of Chapter 5 on methods and methodological rules from Erduran and Dagher's (2014) book. In Chap. 2 of this book, we have illustrated Brandon's (1994) analysis of methods. Although Brandon explores the idea of experiment in the context of biology, his work is relevant for other science domains such as chemistry. Brandon (1994) depicts two ways in which experiments are usually contrasted: contrast with observations and contrast with descriptive work. Critical to the contrast between experiment and observation is the occurrence of manipulation. In terms of the contrast of experiment with descriptive work, a key factor to the contrast is whether a hypothesis is being tested or whether the values of parameters are being measured. Parameter measures may demand considerable manipulation but may or may not involve the testing of hypotheses. Brandon's examples illustrate that not all experiments involve hypothesis testing and that not all descriptive work is non-manipulative. Brandon represents the connections between experiments and observations in terms of a two-by-two table (see Table 4.2). The nature of the investigation (experiment/observation) is related to whether or not (a) it involves manipulation and (b) hypothesis testing or parameter measure as detailed in Chap. 2.

In the session, the pre-service teachers are presented with Brandon's matrix and asked to engage in a discussion. Essentially, they are given an argumentation task where they have to justify or refute a claim about the nature of methods in science. One claim is presented as all science disciplines using the same methods, while the alternative claim presents the position that there are distinct methods in different sciences. Pre-service teachers are then asked to work in groups to generate reasons to support their claims and to produce a poster to summarise the key points of their discussion:

*Claim 1: All science disciplines use the same methods. There is one universal scientific method.*

*Claim 2: All science disciplines do not use the same methods. Each discipline uses a different method.*

- *In your groups, each person will choose either claim 1 or 2 and produce a list of reasons to support this claim. (You may or may not agree with the claim. Focus on producing the reasons, not agreeing or disagreeing with the claim.)*

**Table 4.2**  Experiment/observation, manipulation/non-manipulation and descriptive/experimental categories

| Experiment/ observation | Manipulate | Not manipulate |
|---|---|---|
| Test hypothesis | Manipulative hypothesis test | Non-manipulative hypothesis test |
| Measure parameter | Manipulative description or measure | Non-manipulative description or measure |

Reproduced from Brandon (1994, p. 63)

- *You will then argue your position with the opposite point of view. After discussing your claims in your groups, produce a poster to summarise the key ideas for each claim.*

The task immerses the participants in the application of Brandon's matrix. The visual tool based on a gear analogy proposed by Erduran and Dagher (2014) is also used to supplement the discussion (see Fig. 4.1c). The participants were encouraged to consult pages 92–104 from Erduran and Dagher's (2014) book to produce the reasons to support their claims. The main emphasis in the session, then, is to get pre-service teachers to understand that (a) there are different methods in science, not just one method, and (b) that science works through the use of all of these methods in the collection and evaluation of evidence that contribute to the building of models and theories. Hence, there is a key theme in this session to move beyond consideration of scientific method as a single method of experimental hypothesis testing to a range of methods that also use non-hypothesis testing and descriptive accounts.

### 4.3.5 Session on Scientific Practices

Chapter 4 on Scientific Practices from Erduran and Dagher's (2014) book was assigned as the background reading for the session on practices. The visual tool of scientific practices (i.e. Benzene Ring Heuristic) summarises some key concepts such as data collection, experimentation, modelling and prediction (see Fig. 4.1b). It was introduced to pre-service teachers through an example. The example was about acid-base chemistry and illustrated how, in modelling the chemical reaction of acids and bases at neutralisation, chemists engage in observations and classifications as key activities. They collect data on properties of acids and bases and propose models (e.g. Arrhenius, Brønsted-Lowry) in order to explain the chemical reaction between acids and bases. The particular model explains why neutralisation takes place and it can also be used to predict how other examples of acids and bases might react at neutralisation. The chemical reactions form part of the representation practices. The fact that there are multiple models of acids and bases can be a point for discussion in evaluation which neutralisation reactions can be explained by what mechanism. (For an extended discussion on how the topic of acids and bases can be situated in scientific practices, see Chap. 2.)

Following the introduction of the BRH, the groups were tasked with the discussion of the following statements in their groups:

- *Scientists do not always do experiments. Sometimes they just observe phenomena, and they collect data based on these phenomena. They may classify the data to see trends.*
- *Scientists represent objects or processes, and they engage in discussions while doing science. Representations and discussions do not come at the end of scientific inquiry. They are part of doing science.*

The statements were based on the assumption that scientific practices include a range of cognitive-epistemic and social processes which are interlinked. For instance, modelling of data is mediated by representation and argumentation about various representations. A key assumption in the design of the BRH is that the discursive, representational and epistemic features of scientific practices operate in unison. The discussion points above were intended to communicate some basic ideas about this assumption in the session. The pre-service teachers discussed these points in groups and subsequently presented a summary of their group discussion to the whole class. The session also included a whole class conversation about the pedagogical applications of BRH mediated by the following question: "Do you think that the Benzene Ring Heuristic can be used as a pedagogical tool in science lessons? How?"

### 4.3.6   Session on Scientific Knowledge

The background reading for the Scientific Knowledge session was Chapter 6 from Erduran and Dagher's (2014) book. The key themes promoted in this session were that (a) there are different forms of scientific knowledge; (b) these forms of knowledge are theories, models and laws; (c) there needs to be coherence between the content of theories, models and laws; (d) theories, models and laws operate in unison in producing scientific knowledge; and (e) theories, models and laws grow in time contributing to paradigms and at times there might be paradigm shifts leading to the consideration of new theories, models and laws. Pre-service teachers were provided with Table 4.3 that gives some examples from different sciences and illustrates some of the themes specified above.

Initially pre-service teachers considered the examples and discussed any terminology that they did not understand. The subsequent task asked the participants to work in groups to produce new examples of theories, models and laws as well as examples that are not theories, models and laws. Here the intention was to get the pre-service teachers to think about criteria for defining what counts as a theory, what counts as a model, what counts as a law and what does not. The groups then placed the examples in envelopes, and the envelopes were distributed to other members of the class so that the participants could see more examples. When a group

**Table 4.3**  Theories-Laws-Models (TLM) in different science domains

| Form of knowledge | Domain | | |
|---|---|---|---|
| | Biology | Chemistry | Physics |
| Theory | Genetic theory | Atomic theory | Thermodynamics |
| Law | Inheritance law | Periodic law | Laws of thermodynamics |
| Model | Genes | Atomic model | Heat transfer |
| *TLM explain* | Biological traits | Structure of matter | Heat |

From Erduran and Dagher (2014, p. 114)

received the examples written on cards, their task was to identify them as theories, models and laws and explain why. The "envoy" technique was used to encourage a member of a group to visit another group to discuss her group's ideas. In summary, the task used the following steps:

- *Discuss the examples before moving on to the next steps. What do you notice about the terms?*
- *In your groups, produce a list of six other examples of theories, laws and models from any domain of science (i.e. three sets of TLM).*
- *In your list include four science concepts that are NOT examples of theories, laws and models. The purpose of this set of concepts is to promote discussion about what counts as theories, laws and models and what does not.*
- *In your groups, write down the terms on separate pieces of paper, and place the whole set (three sets of TLM and the four unrelated concepts) in an envelope.*
- *Hand over your envelope to the next group. Take a set from the other group!*
- *In your groups, sort out the other group's cards into TLMs! Put the unrelated concepts separately.*
- *One person from each group visits the group who received their cards and discusses the categories of TLM.*

Following this task, the groups moved onto a discussion of paradigms and paradigm shifts. Here they were given particular topics such as phlogiston and atom, and they were encouraged to conduct research from books and the Internet to explain how the words can be used in chemistry. They were asked to produce a poster to illustrate their understanding of "paradigm and paradigm shift."

### 4.3.7 Session on Images of the Epistemic Core

The session capitalised on Erduran and Dagher's (2014) collective account of images that are intended to promote understanding of NOS through visualisation. The background reading was Chapter 8 from their book. The images associated with different categories of the epistemic core were presented to the pre-service teachers, and they were asked to consider, having now been exposed to the images, how these images could be of used for teaching and learning. In the first part of the task, the pre-service teachers carried out a group discussion by considering the following questions:

- *How are visual tools used in science teaching and learning?*
- *What are the strengths and limitations of using visual tools in science lessons?*
- *If you could add another category or image what would it be? Why?*
- *Can you quickly sketch it?*
- *How would you assess students' understanding of "generative images of science"?*

In the second part of the task, the participants were given the images altogether (see Fig. 4.1) and asked to evaluate them. The task proceeded by asking the participants to consider other potential images and the implications of these images for use in education, such as assessment and lesson planning:

- *Choose an image, and write down the criteria for assessing students' understanding of this image as very good, good, satisfactory and needs improvement. For example, list ideas about what would be evidence of good understanding of aims of chemistry.*
- *The generative images can potentially be used at different levels of schooling and across the school year. What do you think is the "ideal" way of using the images for designing lessons across age groups and across lessons (i.e. in a week, a month, a semester)?*

## 4.4   Lesson Ideas on Chemistry Topics Produced by Pre-service Teachers

The teacher education intervention used various strategies presented such as questioning, group discussion, presentations and the use of visual images to support pre-service teachers' learning of the epistemic core of aims and values, practices, methods and knowledge. In a post-session activity, they were encouraged to work in groups and to produce artefacts to extend their learning. Following each session, the pre-service teachers were given a group task to produce some ideas for lessons that they would develop for teaching in schools. The task was not about producing a full-lesson plan at this stage but rather about applying the key concepts covered in the sessions to a new context in order to consolidate understanding. They were encouraged to use the *Generative Images of Science* (Erduran & Dagher, 2014) to help them think about some examples and to draw any pictures in case their own visual representations helped them to express their ideas. The lesson ideas on the epistemic core categories were produced with different topics such as neutralisation and mixtures, and some preliminary pedagogical strategies were considered by the pre-service teachers. For example, for the case of aims and values, they proposed that the teacher conducts a demonstration. Overall, the lesson ideas were fairly limited in terms of the pedagogical strategies proposed although the pre-service teachers were able to adapt their learning of the epistemic themes to new examples.

In the next section, we provide some examples of the lesson ideas produced by pre-service teachers on scientific aims and values, practices, methods and knowledge. The post-session reflections were encouraged because of a growing amount of research evidence that suggest a role for explicit reflection in changing epistemic beliefs (e.g. Lunn-Brownlee, Schraw, Walker, & Ryan, 2016). Specifically, reflection on one's epistemic beliefs about the nature of knowledge and the process of knowing can support changes in epistemic cognition (Muis, 2007). Because sessions focusing on reflection were part of the intervention, the content of these sessions is included in this chapter although they can also be considered as outcomes of the intervention. The full set of examples ranging from tasks designed by us as teacher educators and the input by the pre-service teachers of their own learning representations tell a more holistic and complete story of the design aspects. In Chaps. 5 and 6, the emphasis will shift more towards implementation through the outcomes observed in pre-service teachers' visual representations and verbal data.

### 4.4.1 Lesson Ideas on Aims and Values

The pre-service teachers used the idea of "aims and values" in science by choosing the example of cleaning agents. They represented the coordination of epistemic, cognitive and social aims and values of science by a lab bench where the process of producing cleaning agents relied on particular epistemic aims such as "objectivity" and cognitive processes such as "critical examination". They represented the social aims by highlighting the utility of the chemistry of cleaning agents and application in daily life of doing laundry (Fig. 4.3).

### 4.4.2 Lesson Ideas on Practices

An example of lesson ideas on practices is illustrated in Fig. 4.4. Here the pre-service teachers have used the analogy of the solar system to illustrate scientific practices similar to the Benzene Ring Heuristic. They chose the topic of mixtures to illustrate the different components of the Benzene Ring Heuristic. In their description of the lesson idea, they underlined the key aspects of the heuristic, and they drew a picture of the solar system.

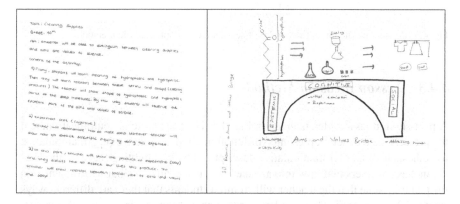

**Fig. 4.3** Pre-service chemistry teachers' post-session group ideas about aims and values of science

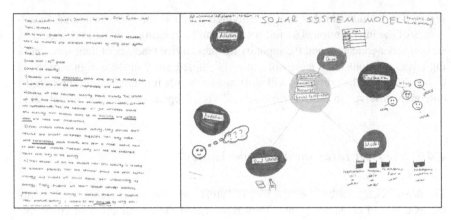

**Fig. 4.4** Pre-service chemistry teachers' post-session ideas about scientific practices

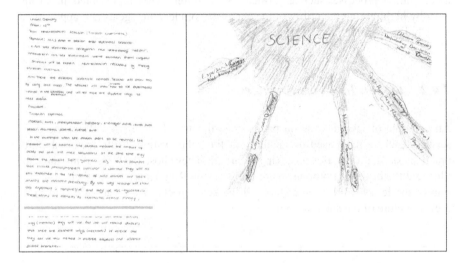

**Fig. 4.5** Pre-service chemistry teachers' post-session ideas about scientific methods

### 4.4.3   Lesson Ideas on Methods

A post-session lesson idea is given in Fig. 4.5. Here the topic is neutralisation, and the pre-service teachers represented different methods in science (e.g. manipulative hypothesis testing, non-manipulative description) by highlighting these methods as "branches" of rivers that flow into a "lake" of science, all contributing to science. In the text they state that the teacher will "remind students that there are different ways (methods) in science". They use keywords such as "experiment", "hypothesis testing" and "manipulative" in the text as well as the picture.

### 4.4.4 Lesson Ideas on Knowledge

Lesson ideas on scientific knowledge included a picture of a bird bath (see Fig. 4.6). In the text, they stated in Turkish that "in illustrating growth of scientific knowledge, the lesson examines the changes and developments in concepts and theories in the context of the atom". They referred to different models of the atom such as Dalton, Thompson, Rutherford and Bohr and likened these models to different levels of the bird bath. Each level was represented as TLM (standing for theories, laws and models), and the most up to date knowledge was analogous to the bottom of the bird bath with a bigger area where more water accumulated. As previously stated, the pre-service teachers were encouraged to use resources in producing further ideas as they reflected on the sessions. It is likely that they obtained the different model accounts from these sources. The inclusion of historical accounts was not an objective of the knowledge session, considering the demands of the existing content of the session. However, it is encouraging that pre-service teachers chose to situate their ideas in a historical context.

In summary, the pre-service teachers were exposed to the theoretical background on epistemic concepts through readings from Erduran and Dagher's (2014) book and were engaged in active learning tasks in the sessions. The tasks involved extended group discussions and production of artefacts including posters and presentations. The use of visual representations was encouraged, and some of the tasks were based on an argumentation framework where the pre-service teachers were deliberately positioned to debate. The post-session task was meant to encourage the pre-service teachers about the key ideas from the sessions and to extend their learning to some lesson ideas that they could potentially use in their own teaching. In all

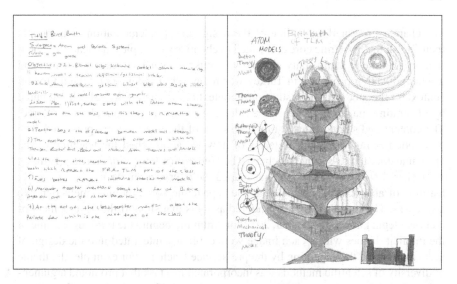

**Fig. 4.6** Pre-service chemistry teachers' post-session ideas about scientific knowledge

categories, the pre-service teachers produced pictures where they extended their understanding to an example analogy. These representations suggest that the pre-service teachers paid attention to the key themes that were promoted in the sessions. For example, in the aims and values session, there was a strong emphasis on the tight connection between the cognitive, epistemic and social aims and values of science. The representation of the laboratory bench as a continuous space seems to echo this point.

Often the pre-service teachers' drawings included analogies to chemistry as well, including in this case with the reference to a laboratory bench. In the case of scientific practices, the theme of connectedness of the various components was highlighted through language such as "all distance to sun is the same" (Fig. 4.5). Although the statement itself of course is not scientifically true, it serves the purpose of communicating how scientific practices occur in unison in the operation of science. The scientific method picture illustrates how science utilises different methods, and it is the coordination of evidence obtained through these methods that leads to scientific understanding. Finally the idea of growth in scientific knowledge is expressed in the bird bath analogy where the levels of the bath get bigger and bigger and more water accumulates as one goes down the bath. Overall, the pre-service teachers applied their learning from the sessions to the lesson ideas they produced in the subsequent session, and their pictures suggest that they were mindful of the key themes from each session. Their descriptions highlight how they can imagine teaching aims and values, practices, methods and knowledge in the context of chemistry as they made explicit links to chemistry in the analogies that they drew on.

## 4.5    Conclusions

The chapter presented an outline of the design and implementation of a module focusing on the epistemic core embedded in chemistry examples in a teacher education programme at a university in Turkey. Status of science education research and teacher education provision in Turkey was reviewed in order to provide a broader context for the module. The institutional context of the module was also highlighted by illustrating the particular university department where the module was offered. The content and strategies used in the sessions of the module were presented. The description is fairly specific in terms of how we approached the teaching of the epistemic core to pre-service teachers. The *Generative Images of Science* (Erduran & Dagher, 2014) were key tools that shaped the intervention. As noted in Chap. 2, learning of abstract concepts such as the "epistemic core" requires metacognitive awareness. The images can be considered as tools that support the development of meta-strategic knowledge (Zohar, 2012) or "thinking behind the thinking". Some of the implicit themes within each image were explicitly integrated into the design of tasks to promote their learning by the pre-service teachers. For example, the theme of diversity of scientific methods was incorporated in a task that promoted argumentation about alternative claims on the nature of scientific methods. In all sessions, argumentation and visual tools were used to complement other strategies such as group discussions, presentations and questioning. The module dedicated time to the extension of the content of each session to the transfer of learning to a new topic where the pre-service teachers developed their own analogies and representations to adapt the content of the sessions.

Ultimately the intervention aimed to influence pre-service teachers' epistemological perspectives in general and chemistry in particular. As Tsai (2006) illustrated, science teachers who hold dualist views of science knowledge directed their students' attention to test scores and devoted more instructional time and efforts on lectures, tutorials and drilling. They tended to focus on the absolute correctness of knowledge and facts and rely on the authority of teachers and textbooks (Tsai & Kuo, 2008). Mansour (2013) reported that science teachers who viewed science knowledge as valid, absolute and cumulative consistently viewed teaching as conveying knowledge to learners. They perceived learning as the passive accumulation of knowledge. The pre-service teachers in our project were exposed to a coordinated approach to considerations of the nature of knowledge and processes of knowledge production. In other words, they systematically covered, through purposeful tasks, the means (e.g. practices and methods), the reasons (i.e. aims and values) and the outcomes (i.e. knowledge) of scientific inquiry. They considered the epistemic core in the context of their subject domain of chemistry and, as such, produced analogies that drew on chemistry examples. Their discussions culminated in the production of group posters that reflected the consensus of the group's ideas.

Due to the timing of the module delivery in the teacher education programme as well as constraints regarding the amount of time available for pre-service teachers to engage in teaching practice, it was not possible to follow them up in schools to observe their teaching practice. Ideally, the ideas produced by pre-service teachers could be extended into lesson plans, and the pre-service teachers could teach lessons based on these lesson plans that have direct links to the themes covered in the sessions. Subsequent feedback on teaching practice could consolidate the development of further understanding. While the intervention project allowed us to gain some insight into how pre-service teachers might deal with themes such as the epistemic core of chemistry, it did not provide any indication of pre-service teachers' pedagogical content knowledge (PCK) related to them. Shulman's (1986) notion of PCK describes the kind of understanding and knowledge that teachers need to have in order to teach. He described PCK as "The most useful forms of content representation... the most powerful analogies, illustrations, examples, explanations, and demonstrations—in a word, the ways of representing and formulating the subject that makes it comprehensible for others" (p. 9). The pre-service teachers' adaptation of the visual images to new examples coupled with understanding of the various categories of the epistemic core may be the first step in the development of their PCK.

The description of the design and the impact of the intervention with examples of pre-service teachers' outputs illustrate what is possible to accomplish in teacher education, in particular in fostering epistemic thinking. Overall, the chapter illustrates the content of the teacher education sessions including the actual tasks and strategies used in the module making it possible for other teacher educators to adapt and use them in their own teaching of pre-service teachers. In Chap. 5, we present further evidence on pre-service teachers' perceptions of the epistemic themes targeted in the sessions (e.g. defining aims and values, types of scientific practices, diversity of scientific methods as well as coherence among knowledge forms and growth of scientific knowledge) by reviewing in more depth their individually produced drawings and verbal data from interviews.

# References

Akkus, H., & Uner, S. (2017). The effect of microteaching on pre-service chemistry teachers' teaching experiences. *Cukurova Universitesi Egitim Fakultesi Dergisi, 46*(1), 202–230.

Allchin, D. (2011). Evaluating knowledge of the nature of (whole) science. *Science Education, 95*(3), 518–542.

Baird, D. (2000). Analytical instrumentation and instrumental objectivity. In N. Bhushan & S. Rosenfeld (Eds.), *Of minds and molecules* (pp. 90–114). Oxford, UK: Oxford University Press.

Ben-David, A., & Zohar, A. (2009). Contribution of meta-strategic knowledge to scientific inquiry. *International Journal of Science Education, 31*(12), 1657–1682.

Brandon, R. (1994). Theory and experiment in evolutionary biology. *Synthese, 99*, 59–73.

Brickhouse, N. W., & Bodner, G. M. (1992). The beginning science teacher: Classroom narratives of convictions and constraints. *Journal of Research in Science Teaching, 29*, 471–485.

Brown, J., Collins, A., & Duguid, P. (1989). Situated cognition and the culture of learning. *Educational Researcher, 18*(1), 32–42.

Cakiroglu, E., & Cakiroglu, J. (2003). Reflections on teacher education in Turkey. *European Journal of Teacher Education, 26*(2), 253–264.

Duschl, R. A., & Erduran, S. (1996). Modeling growth of scientific knowledge. In G. Welford, J. Osborne, & P. Scott (Eds.), *Research in science education in Europe: Current issues and themes* (pp. 153–165). London: Falmer Press.

Erduran, S., & Dagher, Z. R. (2014). *Reconceptualizing the nature of science for science education: Scientific knowledge, practices and other family categories.* Dordrecht, The Netherlands: Springer.

Erduran, S., & Mugaloglu, E. Z. (2016). Trends in science education research in Turkey: A content analysis of key international journals from 1998–2012. In M. H. Chiu (Ed.), *Science education research and practice in Asia: Challenges and opportunities* (pp. 275–288). Dordrecht, The Netherlands: Springer.

Erduran, S., & Kaya, E. (2018). Drawing nature of science in pre-service science teacher education: Epistemic insight through visual representations. *Research in Science Education, 48*(6), 1133–1149.

Eret, E. (2013). An assessment of pre-service teacher education in terms of preparing teacher candidates for teaching. Unpublished PhD dissertation, Middle East Technical University, Turkey.

Gess-Newsome, J. (1999). Secondary teachers' knowledge and beliefs about subject matter and their impact on instruction. In J. Gess-Newsome & N. G. Lederman (Eds.), *Examining pedagogical content knowledge* (pp. 51–94). Dordrecht, The Netherlands: Kluwer Academic Publishers.

Kaya, E., & Erduran, S. (2016). Yeniden Kavramsallaştırılmış "Aile Benzerliği Yaklaşımı": Fen Eğitiminde Bilimin Doğasına Bütünsel Bir Bakış Açısı. *Türk Fen Eğitimi Dergisi, 13*(2), 76–89. ISSN:1304-6020. https://doi.org/10.12973/tused.10180a (In Turkish, Reconceptualized "Family resemblance approach": A holistic perspective on nature of science in science education).

Kaya, E., Erduran, S., Aksoz, B., & Akgun, S. (2019). Reconceptualised family resemblance approach to nature of science in pre-service science teacher education. *International Journal of Science Education, 41*(1), 21–47. https://doi.org/10.1080/09500693.2018.1529447

Kuhn, D. (1999). Metacognitive development. In L. Balter & C. Tamis-LeMonda (Eds.), *Child psychology: Handbook of contemporary issues* (pp. 259–286). Philadelphia: Psychology Press.

Kuhn, D. (2001). Why development does (and does not) occur: Evidence from the domain of inductive reasoning. In J. L. McClelland & R. S. Siegler (Eds.), *Mechanisms of cognitive development: Behavioral and neural perspectives* (pp. 221–249). Mahwah, NJ: Lawrence Erlbaum Associates.

Kuhn, D. (2002). What is scientific thinking and how does it develop? In U. Goswami (Ed.), *Blackwell handbook of childhood cognitive development* (pp. 371–393). Malden, MA: Blackwell Publishing Ltd.

Lave, J., & Wenger, E. (1991). *Situated learning: Legitimate peripheral participation.* Cambridge, UK: Cambridge University Press.

Lederman, N. G., Abd-El-Khalick, F., Bell, R. L., & Schwartz, R. (2002). Views of nature of science questionnaire (VNOS): Toward valid and meaningful assessment of learners' conceptions of nature of science. *Journal of Research in Science Teaching, 39*(6), 497–521.

Loucks-Horsley, S., Hewson, P. W., Love, N., & Stiles, K. E. (1998). *Designing professional development for teachers of science and mathematics.* Thousand Oaks, CA: Corwin Press.

Luft, J. A. (2007). Minding the gap: Needed research on beginning/newly qualified science teachers. *Journal of Research in Science Teaching, 44*(4), 532–537.

Lunn-Brownlee, J., Schraw, G., Walker, S., & Ryan, M. (2016). Changes in preservice teachers' personal epistemologies. In J. A. Greene, W. A. Sandoval, & I. Braten (Eds.), *Handbook of epistemic cognition* (pp. 300–317). New York: Routledge.

Mansour, N. (2013). Consistencies and inconsistencies between science teachers' beliefs and practices. *International Journal of Science Education, 35*, 1230–1275.

Matthews, M. (2012). Changing the focus: From nature of science (NOS) to features of science (FOS). In M. S. Khine (Ed.), *Advances in nature of science research* (pp. 3–26). Dordrecht, The Netherlands: Springer.

Muis, K. R. (2007). The role of epistemic beliefs in self-regulated learning. *Educational Psychologist, 42*, 173–190.

NGSS Lead States. (2013). *Next generation science standards: For states, by states.* Washington, DC: National Academies Press.

Reiser, B. (2013). *What professional development strategies are needed for successful implementation of the Next Generation Science Standards.* K-12 Centre at ETS: International Research Symposium on Science Assessment.

Shulman, L. S. (1986). Those who understand: Knowledge growth in teaching. *Educational Researcher, 15*(2), 4–14.

Supovitz, J. A., & Turner, H. M. (2000). The effects of professional development on science teaching practices and classroom culture. *Journal of Research in Science Teaching, 37*(9), 963–980.

Teo, T. W., Goh, M. T., & Yeo, L. W. (2014). Chemistry education research trends: 2004–2013. *Chemistry Education Research and Practice, 15*, 470–487.

Tsai, C. (2006). Reinterpreting and reconstructing science: Teachers' view changes toward the nature of science by courses of science education. *Teaching and Teacher Education, 22*, 363–375.

Tsai, C., & Kuo, P. (2008). Cram school students' conceptions of learning and learning science in Taiwan. *International Journal of Science Education, 30*, 353–375.

Tumay, H. (2016). Emergence, learning difficulties and misconceptions in chemistry undergraduate students' conceptualizations of acid strength. *Science & Education, 25*, 21–46.

Wenger, E. (1998). *Communities of practice learning, meaning, and identity.* Cambridge, UK: Cambridge University Press.

Wenger, E., McDermott, R., & Snyder, W. (2002). *Cultivating communities of practice: A guide to managing knowledge.* Cambridge, MA: Harvard Business School Press.

YÖK. (2007). *Öğretmen yetiştirme ve eğitim fakülteleri (1982–2007): Öğretmenin üniversitede yetiştirilmesinin değerlendirilmesi*. Ankara, Turkey: Yüksek Öğretim Kurulu Yayını.

YÖK. (2018). *Kimya öğretmenliği lisans programı*. Ankara, Turkey: Yüksek Öğretim Kurulu Yayını.

Zohar, A. (2012). Explicit teaching of metastrategic knowledge: Definitions, students' learning, and teachers' professional development. In A. Zohar & Y. J. Dori (Eds.), *Metacognition in science education: Trends in current research* (pp. 197–223). Dordrecht, The Netherlands: Springer.

Zohar, A., & Ben-David, A. (2008). Explicit teaching of meta-strategic knowledge in authentic classroom situations. *Metacognition Learning, 3*, 59–82.

# Chapter 5
# Pre-service Chemistry Teachers' Representations and Perceptions of the Epistemic Core: A Thematic Analysis

## 5.1 Introduction

The design of the teacher education intervention reported in Chap. 4 was guided by our review of philosophy of chemistry and teacher education literatures reported in Chaps. 1, 2 and 3. The theoretical ideas compiled from these reviews were transformed into teachable content in the teacher education intervention, and they were implemented with pre-service teachers in Turkey. Some of the outputs that pre-service teachers produced after each session were included in Chap. 4. These were generated when the pre-service teachers reflected on the content of the previous session and applied their own ideas using different topics. In this chapter further outcomes are detailed relative to the themes that were used in the sessions, such as diversity of methods and coherence between forms of scientific knowledge. The focus is on example outcomes for the three chemistry pre-service teachers to illustrate the effect of the teacher education intervention. The three pre-service teachers will be referred to with the pseudonyms Zerrin, Berna and Dilek. These pre-service teachers were in a group with another individual named Alev whose case will be described exclusively in Chap. 6.

Zerrin, Berna and Dilek were female pre-service chemistry teachers in the 4-year teacher education programme at Bogazici University in Turkey. They enrolled in the elective course in which teacher education intervention was carried out in their last semester in the programme. Prior to their participation in the intervention, they had already taken courses on content knowledge, pedagogy and electives such as those offered from other departments including the humanities. While these three pre-service chemistry teachers worked together and produced the outcomes reported in subsequent sections together in a group, they were actually quite different in terms of their motivation and productivity throughout the intervention. For example, Zerrin was more motivated, enthusiastic and interested in the sessions when compared to the others. She had the highest GPA in the group whereas Alev had the

© Springer Nature Switzerland AG 2019
S. Erduran, E. Kaya, *Transforming Teacher Education Through the Epistemic Core of Chemistry*, Science: Philosophy, History and Education,
https://doi.org/10.1007/978-3-030-15326-7_5

lowest. Berna's GPA was the second highest and Dilek's was the third highest. Zerrin participated in all sessions and was engaged in the activities. The evidence from her verbal statements and visual representations supported this observation by us as the instructors of the sessions and knew the pre-service teachers fairly well considering that we have spent 11 weeks on the module with them. Given Zerrin's enthusiasm, many of the outcomes to be presented in this chapter are derived from data on her in order to illustrate optimal outcomes of teacher education interventions as intended by design. However, we are also mindful of illustrating what is possible with pre-service teachers who find the content of the intervention challenging or uninteresting. For this reason, we also use examples from Berna and Dilek to give other teacher educators a sense of what potentially can be expected in their own teaching contexts.

Data sources on pre-service teachers included are drawings which illustrate their visual representations of the epistemic core and structured individual interviews which detail their verbal accounts of each category. Our choice of these data sources is deliberate in the sense that we wanted to provide the pre-service teachers with the opportunity to express their thinking through different modes of expression. Verbal and visual expressions used in unison are more likely to capture the nuances of the pre-service teacher's thinking than just one mode of expression alone. There were 11 teacher education sessions as described in Chap. 4 (see Table 4.1 in Chap. 4). It is worthwhile to recount the structure of the intervention in order to contextualise the results. In the teacher education sequence, there was an overlap of the reflection on one category and the introduction of another category. For example, reflecting and thinking about lesson ideas related to Aims and Values on Monday was followed by an introduction to workshop-style activities on Scientific Practices on the Thursday of the same week. All teacher education courses at the university are covered over a 14-week period. It is also typical that each session would be split into two sessions for a particular week. Hence, the particular decisions regarding the placements of the sessions were made on the basis of the overall expectations of the teacher education programme and the requirements within the university. The pre-service teachers were encouraged to bring together all categories in Weeks 9 and 10 to produce projects which they presented during Weeks 12 and 13. The post-data collection occurred in Week 14. The "epistemic core" sessions were covered from Weeks 3 to 7. Hence, there were 7 weeks from the end of the last session on knowledge and the collection of post-intervention data which included drawings and interviews.

## 5.2    Tracing Pre-service Teachers' Representations and Perceptions

As we highlighted in Chaps. 3 and 4, the use of visual representations (e.g. Eilam & Gilbert, 2014) and analogies (Aubusson, Treagust, & Harrison, 2009) were targeted as strategies in the teacher education intervention due to research evidence about

**Table 5.1** Data sources on pre-service teachers' visual representations and perceptions

| Pre-service teacher | Visual representations from drawings | | | | Perceptions from interviews | | | |
|---|---|---|---|---|---|---|---|---|
| | A/V | P | M | K | A/V | P | M | K |
| Zerrin (highest GPA) | x | x | x | x | x | x | | x |
| Berna (second highest GPA) | | | | x | | x | x | |
| Dilek (third highest GPA) | | | | | | | | x |

A/V aims and values, P practices, M methods, K knowledge

their utility in supporting teachers' learning. As the amount of data across these three pre-service teachers are quite large for an in-depth exploration, we selected particular data to focus on given what seemed relevant to illustrate the observed outcomes relative to the intended outcomes. The data sources are illustrated in Table 5.1 where x indicates the data used for a particular pre-service teacher.

For each category of the epistemic core (i.e. aims and values, practices, methods and knowledge), Zerrin's drawings are used. This is because within the group, Zerrin's drawings captured all the intended instructional goals, and we want to illustrate them in a coherent manner. The selection of the episodes from the interviews was based on the quality of content relative to the intended instructional goals of each epistemic core category. Zerrin's interview for the methods category was fairly limited in this sense. Therefore, Berna's interview is used to for this purpose. Berna's motivation and engagement can be interpreted as medium within the group. She participated in almost every session and engaged in the activities although not as enthusiastically as Zerrin did. Berna's perceptions of scientific methods and practices are presented along with her drawing of scientific knowledge. Among all the epistemic core categories, the session on scientific knowledge was the most challenging for the pre-service teachers. For this reason, we include a drawing that captures how an average performing pre-service teacher might interpret visually the themes covered in this session. On the other hand, Dilek was the least enthusiastic pre-service teacher in the group. She did not engage fully in discussions. Her verbal statements from the individual interview were also fairly limited, and hence, only her perceptions regarding scientific knowledge are presented since the other data sources did not yield useful data that shed light on the learning outcomes intended by the teacher education intervention.

In order to investigate the effect of the teacher education intervention on pre-service teachers' understanding of the epistemic core, their representations and perceptions were analysed. Pre-service teachers' representations of epistemic core of science were examined by an instrument in which pre-service teachers were asked to draw pictures to communicate aims and values, practices, methods and knowledge in the context of their subject matter, chemistry, and explain what they drew. So although the questions were kept fairly broad in terms of encouraging them to think scientifically, given their subject specialism of chemistry, they were encouraged to contextualise their responses with chemistry examples. Pre-service teachers' perceptions were assessed by interviewing them individually after the intervention. Their drawings were investigated by considering the aims of the workshop for each category of the epistemic core. For example, in the practices session, the Benzene Ring Heuristic (BRH) was used as a tool to teach scientific practices and the relationships among these scientific practices. Therefore, pre-service teachers' drawings of scientific practices were evaluated in terms of presence of all scientific practices as framed by BRH. The pre-service teachers' drawings of scientific methods were evaluated by considering different scientific methods since the scientific method session focused on Brandon's matrix (Brandon, 1994) which presents the diversity of methods such as manipulative versus non-manipulative hypothesis testing.

The pre-service teachers' perceptions of aims and values, scientific practices, scientific methods and scientific knowledge were identified through interviews after the intervention to understand the overall impact of the intervention on their perceptions. The interview questions were "What comes to your mind when you hear scientific practices" and "Do you think scientific practices are taught in science lessons? If yes, how are scientific practices taught? If not, how can scientific practices be taught in science lessons?" In other words, the interview questions targeted not only their understanding of the epistemic core of chemistry but also their views on how each category could be promoted in teaching and learning. The interview data were analysed qualitatively to investigate how the intervention influenced pre-service teachers' perceptions of the epistemic core. In order to detail the impact of the intervention, we illustrate one pre-service chemistry teacher's (i.e. Zerrin's) visual representations of that category and the perceptions of Berna and Dilek derived from the verbal data. In the subsequent sections, the purpose is to illustrate the outcome of the pre-service teachers' learning given the input from the sessions. In Chap. 6, the focus will be on one pre-service teacher, Alev, whose pre- and post-intervention representations and perceptions will be discussed in depth in order to trace the impact of the intervention.

## 5.3  Defining Aims and Values of Science

In the teacher education intervention, the aims and values workshop highlighted the epistemic, cognitive and social aims and values of science. A triangle representing three components of aims and values of science developed by Erduran and Dagher (2014) (see Chap. 4) was used as a tool to facilitate pre-service teachers' differentiation of these different kinds of aims and values. After the teacher education intervention, Zerrin drew a pine tree and wrote *epistemic*, *cognitive* and *social* on this drawing to represent aims and values of science (see Fig. 5.1). In her explanation of this representation, she mentioned that "Aims and values has three important parts: social, epistemic and cognitive". Then she explained that "Social parts are related to society's ethos like honesty. Epistemic and cognitive parts are related to knowledge and using this knowledge". The pine tree that Zerrin drew, which is very similar to the triangle visual of aims and values, can be interpreted as evidence of adaptation

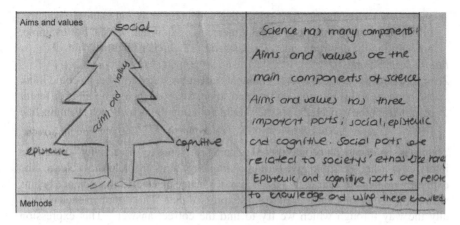

Aims and values

social

aim and values

epistemic

cognitive

Science has many components! Aims and values oe the main components of science. Aims and value) ha) three important ports; social, epistemic and cognitive. Social ports are related to societys' ethos the tone Epistemic and cognitive ports oe relate to knowledge and using these knowled

Methods

**Fig. 5.1** Zerrin's drawing of aims and values of science

of the aims and values of science triangle used in the session. Zerrin also could emphasise three components of aims and values of science which are epistemic, cognitive and social aims and values. In addition to her drawing and explanation, her examples for each component of aims and values such as honesty suggest that the input from the session was memorable to Zerrin.

In the interview conducted after the intervention, Zerrin mentioned what came to her mind about aims and values of science. First, she referred to "serving humanity" as an aim of chemistry. Then she explained cognitive, epistemic and social aims and values of science and gave some examples. For example, she stated that cognitive aims include methodological rules and production of scientific knowledge as well as ethical issues like "not giving harm to the environment" as the social dimension of aims and values of science. Furthermore, she mentioned that scientists should not consider their personal ideas related to religion or culture while doing their science. This expression might be interpreted as "objectivity" which was one of the epistemic-cognitive aims and values of science promoted in the teacher education session. Overall, Zerrin could define and give specific examples of epistemic, cognitive and social aims and values of science.

The interview data related to defining aims and values of science from Zerrin's post-interview are presented below. (We should note that since Turkish – the original language of the verbal data – is gender neutral, there was no distinction between references to personal pronouns. We have used "he" throughout the book for simplicity's sake, as there is no direct translation of the gender-neutral personal pronoun in English. By use of "he/him", we are referring to the gender-neutral meaning of the pronoun in Turkish.)

*Interviewer:* What comes to your mind when I say aims and values of science?
*Zerrin:* First of all, it should serve humanity. So in general, my thoughts are about making life easier. In my opinion, it should answer some questions. Through the lesson, we divided aims and values of science as cognitive, epistemic and social. For example, cognitive part which includes methodological rules and the process

about how we reach scientific knowledge which are aims and values of science. Ethical issues also consist of social dimension of aims and values. For example, not harming the environment… Or, instead of researcher scientist…

*Interviewer:* Scientist.

*Zerrin:* Scientists should not let his thoughts affect his study. Maybe, he could think the opposite to his religion, culture and so on. So, even though he finds different results from the expected ones, he should not cling to his own ideas. Things like that.

The latter sentences can be interpreted as being referencs to "objectivity". When the interviewer asked Zerrin to give further examples of aims and values of science, she explicitly referred to "curiosity" as an aim of science and stated that science begins with asking questions. Furthermore, she expressed that "The aim of science is actually the way through which we try to find the correct answer". This expression might be interpreted as "accuracy" as an aim of science. Zerrin also referred to "honesty" as a value, and she explained that scientists' religion and sociocultural background should not be a factor that affects what they do or the way they do. Hence, Zerrin added "curiosity" and "accuracy" to her previous examples of "honesty" and "serving humanity":

*Interviewer:* So, could you give some examples of aims and values of science?

*Zerrin:* The aim of science is actually to satisfy people themselves, make themselves happy and satisfy their feelings of curiosity. Science is already emerging in that form. When a person begins to ask questions, then science has already started. It's not just scientists doing science. Students can also do science at school, at the park or anywhere. As I said as before, values of science can be… for example, if a person launches a medicine for human health, he should care about whether or not this medicine has side effects.

*Interviewer:* What are the aims of science?

*Zerrin:* The aims of science… The aim of science is actually the way which trying to find the correct answer. So, the aim is to make the process easier. For example, it could be a process of getting information, doing experiments, and so on. I think all these are included.

*Interviewer:* So, what are the values of science?

*Zerrin:* When you say the values of science, first of all honesty comes to my mind. There are some topics that scientists should care about in relation to humans. As I said as before, religion, socio-cultural background of scientists should not affect his study. The values of science are like that.

In the interviews conducted after the intervention, apart from the pre-service teachers' perceptions of each category of the epistemic core, their perceptions about pedagogical aspects related to these categories were also investigated. For example, when the interviewer asked Zerrin, "Do you think aims and values of science are taught in science lessons?" she replied "no". She mentioned that she did not experience anything related to aims and values of science in her own education in either science lessons in primary school or chemistry, physics and biology lessons in high school. When Zerrin was asked about the ways in which aims and values of science can be taught, she mentioned that the instructor from the sessions could act as a model and emphasised different strategies such as questioning, visualising, group work, constructing posters and homework used during the intervention. Zerrin considered the collaboration between teacher and student to teach aims and values of science in science classroom. She also stated that students should be encouraged to think about these issues and that aims and values of science should be integrated with science topics at the beginning of lessons. Hence Zerrin referred to the limitations of science teaching in terms of the exclusion of aims and values of science, and she suggested the use of some strategies (e.g. visualisation, discussion) which were modelled to them during the teacher education intervention. The quotation about the pedagogical issues raised by Zerrin from the post-interview is given below:

*Interviewer:* Do you think aims and values of science are taught in science lessons?

*Zerrin:* No, they are not.

*Interviewer:*So, how can aims and values of science be taught in science lessons?

*Zerrin:* Actually, they can do it like Professor Ebru. Instructor-student together… Teachers should ask questions to students. The teacher should place the aims and values of science into the science topic at the beginning of the lesson. It is actually the work of the teachers. Question-answer can be one of the methods. The teacher should enable students to explore themselves by working in groups. Actually all these methods are familiar to us so we do not think about it. Students should be encouraged to think. Maybe, the last step can be visualizing the process. I think students can prepare posters or do this as homework.

*Interviewer:* What can be added to science lessons in order to teach aims and values of science?

*Zerrin:* Many things can be included because I did not remember any science lessons from primary school. In high school, science was divided into physics, chemistry, and biology. In my opinion, I also did not remember any talk about aims of science in high school.

Zerrin's drawing and interview excerpts after the teacher education intervention illustrate that she incorporated some themes related to aims and values of science promoted in the session and produced an analogy of a pine tree consisting of cognitive, epistemic and social aims of science. It could be that the triangle figure used in the sessions influenced her drawing of the tree given the similarity in the jagged edges of the tree. She explained aims and values of science by referring to "serving humanity", "not giving harm to environment", "objectivity", "curiosity", "accuracy" and "honesty" as aims and values of science. In her reflections about the pedagogical aspects of aims and values, Zerrin referred to different strategies including questioning, visualising and group work that can be used to teach aims and values of science. Furthermore, she considered what the instructor did during the session. This reference suggests that she was influenced by the pedagogical strategies used in the session, and the instructor's role in modelling the pedagogical approaches such as discussions and visual representations made a mark on this pre-service teacher. Considering that the Aims and Values session was conducted in Week 3 and the interview data were collected in Week 14, it is encouraging to see that the pre-service teacher could still remember the particular concepts as well as the modelling by the instructor after 10 weeks of being exposed to these issues.

## 5.4   Types of Scientific Practices

The scientific practices session covered scientific practices that scientists use while conducting their research. The Benzene Ring Heuristic (BRH) (Erduran & Dagher, 2014) was used as a visual tool to teach scientific practices such as real world, activities, data, model, explanation and prediction (see Chap. 4). The heuristic was introduced in the session with the example of acids and bases, and the pre-service teachers produced further ideas related to other chemistry topics in the reflection session. Reasoning, representation, social certification and dissemination were highlighted as mediational practices. For example, the session covered the example that different models of acids and bases can be contrasted and debated.

After the teacher education intervention, when Zerrin was asked to represent scientific practices, she drew a daisy with six leaves (see Fig. 5.2). In her representation, she wrote six scientific practices such as prediction, explanation, real world, model, activities and data. She also included "representation, reasoning, discourse" in the middle of her drawing. In her explanation, she mentioned all these scientific practices and stated that "they have same importance and related concepts so I chose the daisy model. Each leaf has same importance for daisy". Thus, Zerrin emphasised that scientific practices are equally important, which was an implicit theme in the BRH. BRH positions different scientific practices as being interdependent on each other and not separable. For instance, models cannot be dissociated from the representations used to communicate them.

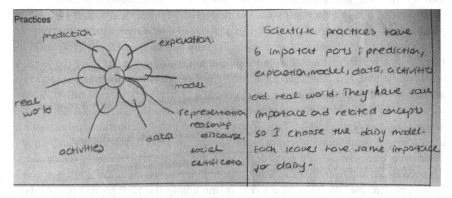

**Fig. 5.2**  Zerrin's drawing of scientific practices

In order to investigate the influence of the session on pre-service teachers' representations and perceptions, the results from both Zerrin's and Berna's interviews are presented here. In her post-interview, when the interviewer asked Zerrin, "what comes to your mind when I say scientific practices?", Zerrin stated "data gathering, observation, and experimenting" as scientific practices. When she was asked to give examples of scientific practices, Zerrin referred to chemistry lessons and mentioned an acid-base reaction. She suggested writing the chemical reaction as a first step and then demonstrating the reaction with an experiment. Hence she used a chemistry context in order to explain what she knows about scientific practices:

*Interviewer:* What comes to your mind when I say scientific practices?
*Zerrin:* Data gathering, observation, these mostly. Experimenting, things like that.
Interviewer:Can you give examples of scientific practices?
*Zerrin:* Yes, I am thinking, scientific practices in chemistry lessons… first writing
   the reaction then going to the laboratory, writing the acid-base reaction and
   showing that it is neutral titration reaction and actually showing with an experiment. Or the other thing could be first. As we go through other questions it will
   cover all but…

In her response, Zerrin continued by referring to prediction, data, observation and experimentation as scientific practices and stated the process as given in the following:

*Zerrin:* First, student can be asked to predict like what will happen at the end of the
   experiment. Then student predicts what is formed at the end. An assignment can
   be given and it starts with prediction. Then everything that they will do would be
   written up step by step. After doing the experiment, the result part, prediction
   part… then they would gather data. These all can be taken as scientific
   practices.
*Interviewer:* What are scientific practices? Can you name them?
*Zerrin:* Do you mean like process? Scientific process, the methodologies and stuff.
   Are we talking about them? When we say scientific practices… I think experiments, observation, data gathering, forming a hypothesis. These approaches.

Even though the question in the above episode was about the definition of scientific practices, the response was more pedagogical in nature. Subsequently the interviewer referred to the pedagogical issues such as teaching scientific practices explicitly and directly asked Zerrin whether or not she thinks scientific practices are taught in science lessons. Zerrin gave an answer by making a comparison between aims and values of science and stated that scientific practices are taught in science lessons superficially when compared to aims and values even if they are not specified as scientific practices. She also gave an example from a biology lesson on examining the structure of the eye. The interviewer then asked Zerrin to give suggestions about teaching scientific practices and what can be included in lessons. Zerrin suggested teaching scientific practices step by step. While she noted that there is no separation among scientific practices in her drawing, she referred to a step-by-step process during the post-interview. Additionally, she noted that some activities are carried out in science lessons but the students do not understand them. She also suggested that teaching scientific practices can either be as a separate subject or as integrated into science lessons. However, she expressed the view that she did not know what could be added to science lessons to teach scientific practices. The excerpt related to teaching scientific practices from Zerrin's post-interview is given below. (Eid al-Fitr is a Muslim holiday when sheep and cows are sacrificed for religious reasons.)

*Interviewer:* Do you think scientific practices are being taught?

*Zerrin:* When you compare to the aims of science, it is mentioned superficially. But not as scientific practices. Maybe before in biology lessons, after Eid al-Fitr the teacher would ask the students to bring in an eye. They would bring the eye and examine it. At least you can carry out an experiment there. It is a practice too but at the end it is not mentioned as a scientific practice. Teaching it is the missing part.

*Interviewer:* Okay, how could scientific practices be taught in science lessons, what could be included in lessons for this aim?

*Zerrin:* Maybe in the things that we do step by step, like we did here… maybe we are doing something in haste so we never know what is that we do. I don't think scientific practices ever take place as a subject matter. I think there is no subject like that. Maybe it can be added, even if it can't be added, it should at least be given to the student with the course of the lesson and its content. Maybe they think they show it but we never see it. How and what could be added to scientific practices… I don't know what could be added.

In summary, although Zerrin was now versed in scientific practices and could define them, because she herself had limited access to teaching, she was not clear about how to teach scientific practices. On the other hand, Berna, when asked what comes to her mind when we say scientific practices, identified observation and experiment as scientific practices. Then she explained experiment by stating manipulative and non-manipulative as different methodologies and said that she is reminded of methodologies. Berna mixed the terminology of practices with scientific methods in a similar manner as she did in the context of the scientific methods part of the interview. (Although scientific practices and methods were covered in different sessions and discussed separately in theoretical accounts, these two categories are actually related as they reflect two aspects of an overall approach to doing science. Practices are about the activities and processes of science such as modelling and predictions, whereas methods emphasise the methodological approaches that underpin the practices, for instance, whether or not there was manipulation or measurement of variables.) When the interviewer asked Berna to give examples, she mentioned the soap experiment and the shrinkage experiment to cover the volume-pressure relationship in the topic of gas laws as examples from chemistry. Then she referred to the lesson ideas that her group produced after the scientific practices session. At the end of the interview, she focused on observation, data, experiment, explanation and modelling as scientific practices:

*Interviewer:* Okay. What comes to your mind when I say scientific practices?

*Berna:* When you say scientific practices, I think of the things obtained by observing, observation. Other than that experiment which can be manipulative or nonmanipulative. Now, at the end of the lesson, I can think about methodologies.

*Interviewer:* Can you give examples of scientific practices?

*Berna:* For example, soap experiment or a simple shrinkage experiment for different gas laws. I did a similar thing while preparing the second lesson plan. We can make them work on volume-pressure relationship as a basic experiment with practices.

*Interviewer:* What do you think scientific practices are?

*Berna:* What are scientific practices... the things we do, observation, data gathering, experimentation... what else, explanation can be included. There is modelling again. This is all that comes to my mind at the moment.

Overall, Zerrin's drawing and post-interview data of Zerrin and Berna suggest that the teacher education intervention influenced their perceptions of scientific practices. They both referred to terminology used in the sessions, and they could link this terminology to thinking about lesson activities based on everyday contexts (e.g. Zerrin's reference to the daisy as an analogy and the eye experiment in the context of religious holidays). Zerrin drew a daisy with six leaves and wrote six scientific practices including model, data and prediction on these leaves. She also wrote representation, reasoning and discourse on the centre of the daisy just like in the Benzene Ring Heuristic (BRH) used to present scientific practices during the session. Zerrin could represent all scientific practices in her drawing after the intervention. When Zerrin's and Berna's ideas about scientific practices after the intervention are examined, it is seen that they expressed the types of scientific practices introduced by the teacher education sessions. For example, Zerrin mentioned three scientific practices which are related to data, prediction and activities such as experiment and observation in her post-interview. She included all scientific practices in her drawing. On the other hand, Berna focused on four scientific practices which are related to explanation, data, model and activities such as experiment and observation in her post-interview. There was no reference to "real world" (a part of the BRH) and the mediational scientific practices (i.e. in the centre of the BRH such as discourse, argumentation, representation and reasoning) in the interviews with either Zerrin or Berna. However, Zerrin's link of her lesson ideas to everyday examples suggests that she was implicitly conscious of linking scientific practices to the real world. In Berna's case, this issue would need to be further unpacked through direct questioning. Overall, data from both pre-service teachers' suggest that after about a 10-week gap following their introduction to scientific practices, they were still able to retain the key concepts covered in the session.

## 5.5   Diversity of Scientific Methods

The epistemic core category of "scientific methods" was covered in the session using a version of Brandon's matrix (Brandon, 1994) which presents the diversity of methods such as manipulative versus non-manipulative and experimentation versus descriptive observations (see Chaps. 2 and 4). In the session, the different methods in science were specified by referring to different fields of science such as chemistry, biology and physics. The particular emphasis of the session was about the diversity of scientific methods based on the aims of different fields. The gear representation developed by Erduran and Dagher (2014) was used as a tool to teach different scientific methods and how they need to be coordinated to use evidence from different sources in order to reach conclusions. In other words, it was stressed to the preservice teachers that scientists do not just employ one method, for instance, the experimental method, but rather a range of methods some of which are descriptive. The example of astronomy was given as a science domain that uses many observational non-manipulative descriptions of historical data. In contrast some problems in chemistry and physics may rely on control of variables through an experimental design where there is direct manipulation variables. After the teacher education intervention when Zerrin was asked to represent scientific methods, she drew a pair of glasses and wrote that "there are many types of glasses such as rift swim mask, sunglasses, prescription glasses" (see Fig. 5.3). She drew an analogy between *types* of scientific methods and types of glasses. Furthermore, she noted a similarity about the *function* of methods in science and the function of glasses for people, by stating that people use glasses to see their environment. Similarly, scientists do not use one way of doing science. They rely on many methods. Thus, Zerrin emphasised diversity of methods used in science by using the everyday example of glasses as an analogy to represent scientific methods.

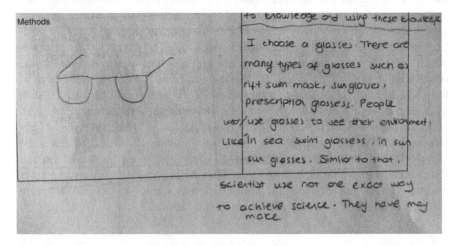

**Fig. 5.3**  Zerrin's drawing of scientific methods

To show the impact of the teacher education intervention on pre-service perceptions of scientific methods, the analysis of Berna's interview is presented here. In the post-interview, when the interviewer asked Berna, "what comes to your mind when I say scientific methods?", Berna first said that she is not sure and then referred to experimentation and observation as scientific methods. Then she continued her explanation by making an explicit reference to the terms *manipulative* and *non-manipulative* which were used in the session. It should be noted that Berna was initially rather sceptical about the entire course. In fact, she had made an appointment with the instructor to query the content of the course as she did not feel that she would be able to manage the content. She indicated that she was taking the course only because she had to take it and that this was the only elective she could take. In light of such background information about her lack of motivation about the module, the fact that she was able to recall some of the key terminology was encouraging:

*Interviewer:* What comes to your mind when I say scientific methods?
*Berna:* Umm.. Scientific methods in science... Actually, I am not sure. It could be more about experimentation, observation.
*Interviewer:* Are these the only two that come to your mind?
*Berna:* We can collect data through observations or making experiments, by inferences. These vary from manipulative or non-manipulative. Things like that.

When the interviewer asked Berna to give some example methods used in science, she focused on social studies and mentioned observation as an example method used in social studies. Subsequently, she emphasised that there are different methods used in science. She also stated experiments as the most used method in different fields such as biology, chemistry and physics. Then she referred to gathering data, observation and model, making inferences as different methods in science. Many of the terms she referred to were covered more explicitly in the scientific practices session. For example, the terms "data" and "model" were covered as part of scientific practices and not as part of the scientific methods session. Although of course there is overlap and these concepts are used in different senses in relation to the key session objectives, the nuance in the coverage of particular themes in each session was different. In the scientific practices session, the Benzene Ring Heuristic (Erduran & Dagher, 2014) dominated the conversation, whereas in the scientific methods, the primary focus was on diversity of methods so a meta-level exploration of scientific practices was promoted. In other words, even though scientific practices and methods are clearly related, the latter was intended to be a more meta-level interrogation of methods in science rather than particular practices that underpin the methodology. In her interview, Berna classified scientific practices as scientific methods. However despite the mixing of the terminology regarding methods or practices, she still focused on the idea of diversity of scientific methods. The quotation related to her depiction of scientific methods is given below:

*Interviewer:* Could you give some examples for scientific methods used in science?

*Berna:* Umm, scientific methods used in science… We can obtain by observations. Especially, social studies are based on observations. So by looking at results from them, or results of surveys…

*Interviewer:* Which methods do you think are used in science?

*Berna:* Which methods… There are many different methods used in science. Maybe, experiments are used mostly for the fields of biology, chemistry, and physics. Gathering data by experiments… Actually here, there are also observations and models. That means making inferences. Like different methods are used. On the one hand, there are observations.

When Berna was asked about whether scientific methods are taught in science lessons, she said that scientific methods are taught as experiments in science lessons. Then, she again referred to social science and mentioned observation as a method used in social science. Berna also expressed that experiment is not the only way for doing science and observations might be mentioned in science lessons in addition to experiment as scientific methods. It seems that Berna needed to consider social sciences when trying to differentiate different methods. First, she classified observation as a method used in social science and experiment as method used in physical/natural sciences. After this differentiation she could reason that experiment is not the only method but that observation could also be used in science. Furthermore, when asked to give specific examples for teaching scientific methods, Berna mentioned theory, law and model as the concepts related to teaching about modelling. To explain the sequence of teaching these concepts, she suggested to start with introducing theory and law to address how they change and then to make students construct a model to indicate the interrelations among theory, law and model. Within the scientific knowledge session, the abbreviation TLM was used, drawn from Erduran and Dagher's (2014) work, standing for theories, laws and models. Berna stated that students could understand how TLM work together and grow scientific knowledge. She also referred to *pool* and *river* as analogies to clarify the concept of scientific knowledge growth. Here Berna mixed the terminology used in the scientific methods and scientific knowledge sessions since she referred to the types of scientific knowledge (i.e. theory, law and model) in order to explain teaching scientific methods. This example is similar to Berna's responses to the scientific methods questions where she used the terminology from the scientific practices session. Here she is also mixing the terminology from the scientific knowledge session when discussing scientific methods. Berna unpacked the pedagogical aspects of teaching scientific methods as follows:

*Interviewer:* Do you think scientific methods are taught in science lessons?

*Berna:* I think they are taught as experiments. Experiments come to mind to answer the questions about what is done in science. However, there is also social science departments. Observations might be mentioned as scientific methods used for social sciences. Hence, for doing science, making experiments are not necessary.

*Interviewer:* Could you give some specific examples to teaching scientific methods?

*Berna:* For instance, theory, law and model concepts can be used when we are teaching about modelling. Firstly, we can mention law and theories to explain how they are changing. These points can be discussed during the lesson and finally, students can construct a model to show the interrelation between theories, law and models. Through this model, students can understand how TLM components support each other and are like a growing mechanism such as a pool or a set of rivers.

As the above quote illustrates, Berna adopted the terminology of TLM and drew an analogy to explain her understanding of it in terms of knowledge growth as well. In other words, she is not only versed in differentiating different types of knowledge, but she took away the idea of the interrelationships (i.e. "how TLM components support each other") and growth of scientific knowledge (i.e. analogy of pool or rivers as well as the explicit use of the word "growth"). These references are in line with the themes intended within the session. However, Berna seems to have appropriated the content of the sessions in a way that would span the terminology used across the different epistemic core categories. In other words, although she seems familiar with the themes promoted in each session, her approach is fairly holistic in using them in her own interpretations. In the context of discussing scientific methods, she is also referring to themes related to scientific knowledge. As previously noted, in the context of discussing scientific practices, she also referred to scientific methods. There is further evidence regarding this point from her post-interview about scientific methods when she was asked to talk about her suggestions for teaching. Here Berna suggested the inclusion of all components of FRA (Family Resemblance Approach) to teach scientific knowledge. As discussed in Chap. 4, there was a whole session dedicated to the FRA where pre-service teachers engaged in activities to justify how we know about science and how science works. FRA was covered as an approach that justifies how we get to know about science. That is, different sciences are classified as "science" because they share certain characteristics (i.e. domain-generality) although they also have differences (i.e. domain-specificity). An activity based on the criteria for characterising biological families was carried out in a session (see Chap. 4). Berna's suggestions for teaching scientific methods included the consideration of issues related to society and ethics as well as aims and values of science:

*Interviewer:* How can scientific methods be taught in science lessons? How can we include this category in science lessons?

*Berna:* I think all of them can be included. Their influence on society and their relation with ethics can also be considered. The methods and aims and values can be also examined. Actually, I think all the components of FRA can be included by the consideration of the topic of the lesson.

It is encouraging to see that Berna was enthusiastic about the use of FRA in science teaching and that her responses in the interview seemed fairly holistic, drawing on all epistemic core categories that were introduced in the sessions. In summary, we have presented Zerrin's drawing and Berna's interview data so far in order to trace pre-service teachers' perceptions about scientific methods. Both Zerrin's representations and Berna's perceptions about scientific methods refer to the diversity of scientific methods. Both pre-service teachers used analogies in their descriptions and explanations. Zerrin drew a pair of glasses to represent the diversity of scientific methods. Berna explicitly referred to manipulative and non-manipulative methods, a distinction that would be very unlikely to have been acquired elsewhere in her education given this is not conventional language in science lessons nor science teacher education. Berna's case was also interesting in the sense that she explained scientific methods in relation to either scientific practices such as experiment, observation, model and data or in relation to scientific knowledge using the terminology of theory, law and model. She also mentioned aims and values in the context of being questioned about scientific methods. Hence, although she mentioned manipulative and non-manipulative methods in science at the beginning of the interview, she progressed to talk about scientific practices and scientific knowledge instead of exclusively focusing on scientific methods. In her consideration of the pedagogical aspects of scientific methods, she brought an even wider frame of reference when she explicitly referred to FRA and suggested the contextualisation of scientific methods in social and ethical contexts.

## 5.6   Coherence Among Knowledge Forms and the Growth of Knowledge

The main goals of the session on Scientific Knowledge were to teach different forms of scientific knowledge which are theory, law and model and how these forms work together to produce scientific knowledge. The session consisted of two activities (see Chap. 4). In the first activity of the session, the pre-service teachers were asked to write theory, law and model (TLM) sets for different topics from chemistry, physics and biology after being presented with some examples. This activity aimed to teach pre-service teachers to differentiate different forms of scientific knowledge. In the second activity of this session, the pre-service teachers were asked to produce a poster reflecting historical development of a specific topic to focus on the idea of "paradigm shift". Furthermore, the visual representation entitled *TLM* developed by Erduran and Dagher (2014) was used. In order to examine the influence of the teacher education intervention on pre-service teachers' representations and perceptions of scientific knowledge, drawings from Zerrin and Berna as well as interview data from Zerrin and Dilek are presented. The scientific knowledge category was the most challenging session for the pre-service teachers based on our observations as teacher educators teaching the session. Hence, two drawings are included to illustrate more examples of how the pre-service teachers interpreted the themes covered in the session.

After the teacher education intervention, when Berna was asked to represent scientific knowledge, she drew a waterfall with fish and wrote TLM to refer to different knowledge forms (see Fig. 5.4). In her drawing, she represented the growth

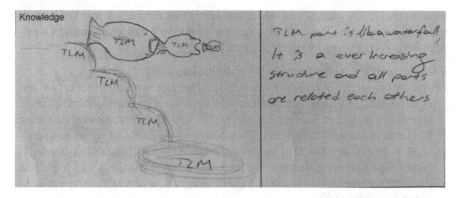

**Fig. 5.4** Berna's drawing of scientific knowledge and growth of scientific knowledge

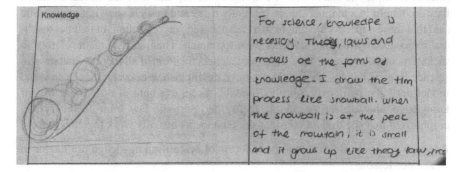

**Fig. 5.5** Zerrin's drawing of scientific knowledge and growth of scientific knowledge

of scientific knowledge by drawing fish in different sizes. She also referred to the relationships among theory, law and model and explained her drawing by writing that "TLM part is like a waterfall. It is an ever increasing structure and all parts are related to each other". Thus, Berna included the key points of (a) forms of scientific knowledge as theory, law and model and (b) the growth of scientific knowledge emphasised in the Scientific Knowledge session. Likewise, Zerrin's drawing included the theme of 'growth' through the use of a snowball analogy and her text reference consisted of an explanation about TLM (Fig. 5.5).

In the post-interview, when the interviewer asked Zerrin about the forms of scientific knowledge, she replied as "theory, law and model". Then she continued her explanation by expressing her previous limited knowledge and misconceptions about scientific knowledge and stated that she was taught only theory and law as scientific knowledge before. She also said that model was not taught in her science lessons. Zerrin mentioned that she was taught if the theory was proven, it would become a law and the laws would never be changed. Then she referred to the teacher education intervention and said that she learnt her previous knowledge about scientific knowledge was wrong and learnt that models are another form of scientific knowledge. The excerpt quotation related to forms of scientific knowledge from Zerrin's post-interview is given below:

*Interviewer:* What are the forms of the scientific knowledge?

*Zerrin:* Forms of the scientific knowledge… The part with the theory, law and model, I guess. I try to remember that what did I say about the forms of scientific knowledge in the first interview, but let me say now. For instance, we didn't learn this issue in our schooling years. They taught us the forms of scientific knowledge as theory and law. The model isn't mentioned with these two forms. If theory is proven, we gather law, but the laws are never changed. So, I learnt that these are not true. Also, I learnt that the model is also a part of this concept and I didn't know that at all.

When Zerrin was asked about teaching different forms of scientific knowledge in science lessons, she stated that she thought theories and laws are taught although they are taught in a manner that is "partially true" and models were not taught. She said that these concepts are taught in a complicated way. She thought that her teachers' expectations from the students were too much. Then, Zerrin again referred to the intervention and said that even if her teachers mentioned all theories and laws in science lessons before, she would not have distinguished between theory, law and model until she participated in this module. As an example of different forms of knowledge, she focused on atomic theory and periodic law from chemistry and emphasised that she did not consider these concepts actually work together before:

*Interviewer:* Do you think that these forms of scientific knowledge are taught in science lessons?

*Zerrin:* I think theories and laws are taught, but they are taught partially true and partially false. The concept of model is not taught. All these concepts are taught in a complicated way and as separate entities. I don't know if they think us to be so clever. I mean, they are mentioning all theories and laws, but I wouldn't separate them as theory-law-model until I took that course. I also didn't separate them in my head, too. For instance, there is an atom theory, there is a law which is a periodic table. But, I didn't consider that these concepts are together.

Subsequently, the interviewer asked Zerrin about her suggestions about the ways to teach scientific knowledge in science lessons. She referred to different fields of science such as biology, physics and chemistry and said that chemistry and physics have more common and united issues. She mentioned that her teacher taught the theories and laws related to the topic of pressure of gases separately by using different ways even if it was a common topic in chemistry and physics. Then she focused on her teacher's high expectations from her students regarding separating and internalising the forms of scientific knowledge. Thus, Zerrin suggested that teachers differentiate theories, laws and models for each topic and then to emphasise the holistic and coherent system which explain that theories, laws and models work together in a systematic way. Zerrin's suggestions about teaching scientific knowledge are given below:

*Interviewer:* How can we teach scientific knowledge in science lessons?

*Zerrin:* How can we teach? I thought that during the course… I mean, we also separated the science as physics, chemistry and biology. But, chemistry and physics have more common and united points. Especially, the topic of pressure of the gases are common in chemistry and physics. But, it is taught us as these laws and theories are separate and they used different ways and names. Actually, they provided us a complicated pool. They expected us to separate and internalize these concepts, but it's really hard. It would be nice to make it easier, such as these are the theories, laws and models of each topic and so on. For instance, they could teach us these entities piece by piece and then they could conclude that they are a whole and coherent system together which are working in a systematic way.

Zerrin continued her suggestions related to teaching scientific knowledge with a snowball analogy to focus the forms of scientific knowledge and the growth of scientific knowledge. The interviewer's question about strategies to teach scientific knowledge and Zerrin's answer to this question are given below:

*Interviewer:* What kind of methods or strategies would you use when you are teaching the forms of scientific knowledge as chemistry teacher?

*Zerrin:* A specific example comes to my mind. In a snowy day, one student makes a small snowball at the top of the mountain. This is actually a piece of knowledge. Then, when the student pushes the snowball on the ground, it becomes bigger and bigger and bigger. Actually, the knowledge is also like this. I mean, theory, law and model are coming together and getting bigger and bigger. It becomes more holistic, broad and assemble. The important thing is that all the three forms of the knowledge are important individually and they also come together to work and progress to provide holistic structure.

As Zerrin's interview data indicate, she could refer to different types of scientific knowledge as well as coherence among these forms of knowledge and growth of scientific knowledge. Likewise, Dilek who was the least motivated pre-service teacher in the group of Zerrin, Berna and Alev, was able to make reference to the knowledge category themes in the post-intervention interview. When Dilek was asked about the forms of scientific knowledge, she explicitly mentioned theories, laws and models as the forms of knowledge and the coherence among them by stating these knowledge forms come together and develop each other. Dilek also referred to the idea of holistic structure of different forms of scientific knowledge. The related quotation from Dilek's post-interview is given below:

*Interviewer:* What are the forms of the scientific knowledge?

*Dilek:* Now, I started to think about this because of our course. The forms of scientific knowledge are based on a theory, law and model. These are coming together and developing each other in time. Then they provide a whole structure. Theories, laws and models.

The interviewer probed Dilek further about whether or not the forms of scientific knowledge are taught in science lessons. Dilek referred to the science curriculum and stated that there is focus related to the difference between theory and law now while there was not such a focus before. She also mentioned that the science curriculum did not include models:

*Interviewer:* Do you think that these forms of scientific knowledge are taught in science lessons?

*Dilek:* I saw the difference between theory and law in the curriculum. I mean, when I was a student I didn't see such kind of thing, but the curriculum has a focus about that issue right now. However, the model wasn't included in the curriculum and we talked that this is a holistic structure. So, I don't think that it is taught. The differences between theory and law are examined only.

When Dilek was asked about the ways to teach scientific knowledge in science lessons, she highlighted using visuals and giving examples from history of science as strategies to teach scientific knowledge. She stated that if scientific knowledge is taught by covering scientists' work, how a theory or a law was formed, students can learn theory, law and model concepts:

*Interviewer:* So, how can we teach scientific knowledge in science lessons?

*Dilek:* I don't know why but I think we can teach the concept with the help of history. I mean, how a theory was formed, how a law was formed and how a related model was provided. All these points can be learnt by the examples, such as which scientist provided what kind of work. I think that kind of approach is more logical to learn the concept.

*Interviewer:* So, how can we include the forms of scientific knowledge in science lessons?

*Dilek:* I think the visuals are pretty important here, again. Because, we are talking about the modelling. Besides that, while we are talking about the theories and laws, we can use visuals and schemes again.

Considering the limited engagement and participation of Dilek in the sessions, it is quite remarkable that after 8 weeks of having been introduced to the idea of TLM, she could retain the key ideas promoted in the session, and she also noted the importance of visual representations, modelling and the use of history of science. Berna drew a waterfall to represent the forms of scientific knowledge (i.e. theory, law and model) and the growth of scientific knowledge which were the main concepts emphasised during the workshop on scientific knowledge. Likewise, Zerrin represented the themes from the session in her drawing about the snowball rolling down a mountain. Furthermore, Zerrin's post-interview data indicate that she could refer to the types of scientific knowledge and the coherence among them to produce scientific knowledge like her drawing of scientific knowledge after the intervention. As for the pedagogical aspect, Zerrin mentioned some misconceptions related to the difference between theory and law and the gap in teaching models as one form of scientific knowledge. Similarly, Dilek's post-interview data show that she could state different types of scientific knowledge which are theory, law and model and the coherence among them. Dilek's perceptions regarding teaching scientific knowledge in science lessons were similar to Zerrin's. Dilek addressed the limited inclusion of model as a form of scientific knowledge in science lessons while theory and law are included somehow in science lessons. Berna, on the other hand, had a way of connecting the themes from different sessions. Considering she took the module not quite willingly and only because she had to take a course for programme requirements, at the end of the module she seemed fairly enthusiastic about reflecting on the content. She not only remembered key themes from each session, but also she was able to communicate a holistic account of the epistemic core categories, linking her ideas for each category across all categories and even referring the FRA as the overall framework underpinning the categories. Overall, the data suggest that the teacher education intervention achieved the intended influence on the group of pre-service teachers discussed in this chapter. It should also be noted that a questionnaire-based pre- and post-test approach including all the pre-service teachers attending the sessions was conducted, and it was found that there was an overall statistically significant difference in favour of the post-test in terms of pre-service teachers' understanding of the FRA categories (Kaya, Erduran, Aksoz & Akgun, 2019). In this book, as we are focusing on an in-depth qualitative investigation into a select number of pre-service teachers, we are not reporting on the other measures of impact that were conducted.

## 5.7  Conclusions

The chapter illustrated the impact of teacher education intervention described in Chap. 4 on pre-service teachers' perceptions of some themes related to the epistemic core. Four chemistry pre-service chemistry teachers worked as a group to complete the tasks and produce the outcomes throughout the intervention. Examples from the learning outcomes of three pre-service chemistry teachers named Zerrin, Berna and Dilek in this group were used to illustrate how the teacher education sessions affected their perceptions of each category of the epistemic core. The representations of the pre-service teachers were examined through their drawings and verbal explanations for each category. Their perceptions of aims and values, methods, practices and knowledge were evaluated through their individual interview data gathered after the intervention. The pre-service teachers' perceptions of each category were presented under different themes related to those categories. These themes were "defining aims and values", "types of scientific practices", "diversity of scientific methods" and "coherence among knowledge forms and the growth of knowledge". In order to indicate the impact of the intervention on pre-service chemistry teachers' representations of each category, Zerrin's drawings after the intervention were used. Pre-service chemistry teachers' perceptions of the epistemic core was reported by focusing on different pre-service teachers' perceptions taken from their post-interview data. For example, Zerrin's perceptions were presented for the defining aims and values section, Berna's for diversity of scientific methods section, Zerrin's and Berna's for types of scientific practices and Zerrin's and Dilek's for coherence among knowledge forms and the growth of knowledge.

Zerrin could refer to the key themes for each category in her explanation of her drawings. For instance, she mentioned cognitive, epistemic and social aims and values, different methods used in science, no hierarchy among scientific practices and growth of scientific knowledge while explaining the picture that she drew. Based on all these examples, it is possible to say that the teacher education intervention influenced Zerrin's representations and perceptions of the epistemic core. For example, she could define aims and values of science including honesty, objectivity, accuracy and so on and understand the types of scientific practices including model, explanation and reasoning. She could verbalise coherence among different forms of the scientific knowledge and the growth of scientific knowledge after the intervention. In fact, observing this progression in her representations and perceptions is not surprising since Zerrin actively engaged in all sessions and was very motivated throughout the intervention. Berna, on the other hand, had an average level of engagement and motivation in the group during the sessions. However, her data suggest that she not only retained many concepts from the sessions but also was able to interrelate these concepts. She explained scientific methods in relation to either scientific practices such as experiment, observation, model and data or in relation to scientific knowledge using the terminology of theory, law and model. She also mentioned aims and values in the context of being questioned about scientific methods. Hence, although she mentioned manipulative and non-manipulative methods in science at the beginning of the interview, she progressed to talk about scientific practices and scientific knowledge instead of exclusively focusing on scientific methods. In her consideration of the pedagogical aspects of scientific methods, she brought an even wider frame of reference when she explicitly referred to FRA and suggested the contextualisation of scientific methods in social and ethical contexts. Dilek as the least motivated pre-service chemistry teacher in the group could differentiate different forms of scientific knowledge as theory, law and model and referred to the idea of growth of scientific knowledge.

The pre-service chemistry teachers' learning outcomes inferred from their drawings and verbal communication suggest a direct impact of the intervention. All pre-service teachers used analogies to represent what they learnt about each category of the epistemic core. They supplemented these analogies with written descriptions. They were able to retain the key themes many weeks after their introduction in the sessions. All three pre-service teachers were keen to draw from pedagogical approaches used in the sessions in making recommendations about teaching the epistemic core. It should be noted that the sessions on the epistemic core were covered in Weeks 4–7, the group was engaged in projects in Weeks 9–10 and the group participated in presentations on their lesson ideas in Weeks 12–13. The group work in Weeks 12–13 in the context of projects most likely consolidated their earlier understandings as they now had to bring together all categories in thinking about lesson activities. Since the data collection on the drawings and interviews happened several weeks after the intervention, it is quite encouraging that the pre-service teachers were able to retain some sophisticated terminology (e.g. manipulative versus non-manipulative) and themes (e.g. interrelationships between knowledge forms).

Overall, the purpose of this chapter was to illustrate what is likely to be observed if the epistemic core ideas are infused in teacher education. The teacher education projects itself was a testing ground for transforming some theoretical ideas for use in teacher education. As such, at this stage, the goal was to investigate how the design of a module focusing on the epistemic core may influence pre-service teachers' epistemic thinking. What was observed is that the pre-service teachers were able to adopt the key themes and concepts introduced in the sessions and they were very creative about using analogies to situate and communicate their understanding. They appealed to chemistry examples in articulating the details of the epistemic core and reflected on the implications for chemistry teaching. Although analogies can be useful pedagogical tools, clearly they also have limitations as widely reported in the research literature (e.g. Aubusson et al., 2009). For example, when a pre-service teacher draws a fish to represent growth of scientific knowledge, she is trying to communicate that each fish corresponds to knowledge at a point in time and that as new evidence emerges, knowledge grows. The bigger fish eats the smaller fish, meaning that knowledge accumulates and grows bigger. However it is possible to have further extrapolations about fish in this scenario and draw out irrelevant or misleading conclusions. For example, there may be inferences about how the disappearance of the small fish may represent loss of knowledge in time. Similarly, the analogy of glasses can be used to stress the diversity of methods relative to the goals of an investigation. However the analogy may result in pre-service teachers drawing some unrelated and unintentional conclusions such as how all scientists may wear glasses.

The pre-service teachers' drawings were used as valuable resources to gain insight into how pre-service teachers interpreted the various themes covered in the sessions. Hence, the drawings reported in the chapter are data originating from individual interviews at the completion of the module. If such drawings were to be integrated into sessions as part of teacher education, such concerns will need to be dealt with by teacher educators. Although analogies can be challenging and they may have limitations, they can also be useful tools for consolidating understanding. In the context of a teacher education session, different analogies can be discussed collectively to precisely deal with the limitations head on so that pre-service teachers develop understanding of the usefulness and shortcomings of analogies. The case of the fourth pre-service teacher, Alev will illustrate further how one particular individual may utilise analogies in communicating her perceptions and understanding of the epistemic core. By using her visual representations and verbal communication before and after the teacher education intervention, Alev's case will be detailed in Chap. 6 to illustrate her progression in light of the intervention. Her case will begin to highlight the developmental aspects of pre-service teachers' learning of epistemic themes. The design of a teacher education programme (see Chap. 4) and its implementation as reported in this chapter are beginning to provide some insights into effective strategies and potential outcomes of including the epistemic core in chemistry teacher education.

# References

Aubusson, P., Treagust D., & Harrison A. (2009). Learning and teaching science with analogies and metaphors. In *The world of science education: Handbook of research in Australasia*. Rotterdam, The Netherlands: Sense Publishers.

Brandon, R. (1994). Theory and experiment in evolutionary biology. *Synthese, 99*, 59–73.

Eilam, B., & Gilbert, J. K. (2014). *Science teachers' use of visual representations*. Dordrecht, The Netherlands: Springer.

Erduran, S., & Dagher, Z. (2014). *Reconceptualizing the nature of science for science education: Scientific knowledge, practices and other family categories*. Dordrecht, The Netherlands: Springer.

Kaya, E., Erduran, S., Aksoz, B., & Akgun, S. (2019). Reconceptualised family resemblance approach to nature of science in pre-service science teacher education. *International Journal of Science Education, 41*(1), 21–47.

# Chapter 6
# The Impact of Teacher Education on Understanding the Epistemic Core: Focusing on One Pre-service Chemistry Teacher

## 6.1 Introduction

The chapter focuses on one pre-service chemistry teacher named Alev who participated in the teacher education intervention described in Chap. 4, and it presents a focused case study (e.g. Hamilton & Corbett-Whittier, 2013; Yin, 2003). Social science researchers use case studies to illustrate details about an individual, group or institution (Miles & Huberman, 1994). There is widespread debate about what counts as a case study, and various distinctions have been made between case studies as a method, methodology and research design (e.g. Hamilton & Corbett-Whittier, 2013). The ultimate purpose of this book is to better understand how pre-service chemistry teachers' engagement in and understanding of the epistemic aspects of chemistry can be supported. We have designed (see Chap. 4) and implemented (see Chap. 5) a teacher education intervention in an effort to investigate how this purpose can be achieved. Considering the lack of research in the broader area of the infusion of philosophy of chemistry and teacher education (see Chaps. 1 and 2), our approach necessitates some fairly detailed description of what is possible to realistically accomplish in teacher education programmes and what is potentially challenging for pre-service teachers' learning. Furthermore, we are interested in how pre-service teachers might interpret abstract epistemic themes in relation to their subject domain and what descriptions (e.g. verbal and visual) they might use in communicating their understandings and perceptions. In this sense, we envisage the use of one pre-service teacher's case as an approach to research, as Elliott and Lukes (2008) argue, that captures the complexity of the appropriation of fairly unfamiliar ideas for pre-service teachers. Data sources on the case study involving the pre-service teacher Alev included in this chapter are (a) drawings which illustrate her visual representations of the epistemic core and (b) structured individual interviews which detail her verbal accounts of each category. Our choice of these data sources is deliberate in the sense that we wanted to provide the pre-service teacher

© Springer Nature Switzerland AG 2019
S. Erduran, E. Kaya, *Transforming Teacher Education Through the Epistemic Core of Chemistry*, Science: Philosophy, History and Education,
https://doi.org/10.1007/978-3-030-15326-7_6

with the opportunity to express her understanding and perceptions through different modes of expression. Verbal and visual expressions used in unison are more likely to capture the nuances of the pre-service teacher's thinking than just one mode of expression alone.

Alev was a member of the group that also included Zerrin, Berna and Dilek whose representations and perceptions were presented in Chap. 5. Alev was the pre-service teacher with the lowest GPA in the group. Alev's pre- and post-representations and perceptions derived from the interview data are compared qualitatively to understand the effect of the intervention on her. As indicated in previous chapters, Alev was a female pre-service chemistry teacher who was in a 4-year teacher education programme. She participated in the intervention during her last semester in the programme. She had already completed many content, pedagogy and elective courses. The details of the programme that she was enrolled in were provided in Chap. 4. It is again worthwhile to recount the structure of the intervention as this has implications in terms of contextualisation of the results. There was an overlap of the reflection on one epistemic core category (i.e. aims and values, practices, methods and knowledge) and the introduction of another category during 1 week across the 11 weeks of the intervention. For example, there was reflection on lesson ideas for aims and values on Monday and workshop-style activities on scientific practices on the Thursday of the same week. All teacher education courses at the university are covered over a 14-week period. It is also typical that each session would be split into two sessions for a particular week. Hence, the particular decisions regarding the placements of the sessions were made on the basis of the overall teacher education provision and the requirements within the university. However, we were mindful of splitting the sessions within a particular week in order to provide periods of pedagogical reflection before introduction of another theme. Hence, we tried to accommodate the goals of reflection and the programme requirements within the design of the teacher education sessions. Finally, the pre-service teachers were encouraged to bring together all categories in Weeks 9 and 10 to produce projects which they presented during Weeks 12 and 13. The post-data collection occurred in Week 14. The "epistemic core" sessions were covered in from Weeks 3 to 7. Hence, there were 7 weeks from the end of the last session on this theme and the conduct of post-intervention data.

Alev was a very outgoing and talkative individual. She was always very vocal in the sessions and engaged in the group activities consistently. In her group with Zerrin, Berna and Dilek, she was the most outspoken and she also participated actively in whole class discussions. In order to assess how Alev visually represented the epistemic core, an instrument was administered before and after the intervention. The instrument required a drawing of what she understands by aims and values, scientific practices, scientific methods and scientific knowledge as well as an explanation of her drawings. Her pre- and post-drawings were analysed by interpreting them relative to the goals of the session on each category of the epistemic core. For example, as reported in Chaps. 4 and 5, one aspect of the scientific knowledge session was the emphasis on theories, laws and models as forms of knowledge. Hence the use of either words or representations that made reference to these concepts was investigated in her drawing of scientific knowledge. In terms of the scientific practices drawings, the Benzene Ring Heuristic (BRH) (Erduran & Dagher, 2014) was used as the main tool to investigate the presence of related concepts in her drawings. In order to assess Alev's perceptions of aims and values, scientific practices, scientific methods and scientific knowledge, she was individually interviewed before and after the intervention. In the interviews, she responded to some questions referring to each category. Samples of the interview questions are "What comes to your mind when you hear aims and values of science and why?" and "Do you think scientific aims and values are taught in science lessons? If yes, how are they taught? If not, how can scientific aims and values be taught in science lessons?" Her pre- and post- interview data were analysed qualitatively to understand the impact of the intervention on her perceptions of the epistemic core.

## 6.2 Representations and Perceptions of Aims and Values

Before the intervention, Alev drew pills and a smiling person who is receiving a treatment with serum, and she stated that she thinks that "aim of science is improve human's life quality". She complemented this drawing with text to explain the reason for her drawing of smiling person by stating that "I draw people who are happy because they are getting healthier" (see Fig. 6.1).

Before the intervention, when the interviewer asked Alev the question "what comes to your mind when I say aims and values of science?", she focused on the improvement of people's lives and gave the example of medicine repeatedly. For her, science should contribute to people's lives, and medicine is one avenue. When pressed to give other and further details, it became clear that she did not really know how to define values of science:

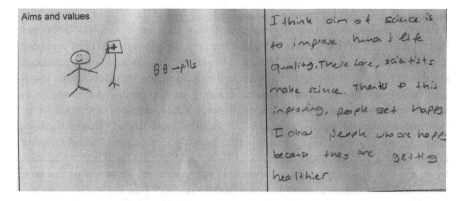

**Fig. 6.1** Alev's drawing of aims and values of science (pre-intervention)

*Interviewer*: What comes to your mind when I say "aims and values of science"?
*Alev*: When you say aims of science, I mean something that facilitate people's lives. In this respect, I think the values of science is also valuable. I think science is valuable because it facilitates and contribute people's lives.
*Interviewer*: So, Could you give any examples for aims and values of science?
*Alev*: Aims and values of science… For instance, we can think of medicine. The aim of medicine is to protect people's health, to heal people's health. In terms of values of science, I think of this because medicine protects people's lives. It is worthwhile.
*Interviewer*: So, regarding the examples you mentioned above, what are you saying when I say the aims of science?
*Alev*: The aims of science… is something that contributes to people's lives. At the same time, it is maybe one of the branches that regulates the nature.
*Interviewer*: So, how can you summarize if I say what are the values of science?
*Alev*: What are the values of science… What do you mean by the value?
*Interviewer*: So, how do you understand? What does it evoke for you when we say values of science?
*Alev*: Actually, nothing comes to my mind when you say values of science.

When the interviewer asked Alev about the pedagogical aspects of aims and values, she indicated that this theme is not something that is included in teaching and learning but that it could be integrated in science lessons because science is already linked to everyday life. She referred to the use of videos as instructional tools:

*Interviewer*: So, do you think aims and values of science are taught in science lessons?
*Alev*: No.
*Interviewer*: So, how can aims and values of science be taught in science lessons?
*Alev*: Firstly, science is interconnected. But we do not learn science like that, namely, in the primary and high school. Science is interconnected with daily life. I think aims and values of science can be taught by giving these types of examples.
*Interviewer*: You mentioned examples. So, what can be integrated into science lessons in order to teach aims and values of science. Which materials can be used?
*Alev*: Daily life examples.
*Interviewer*: For instance, what kind of examples can they be?
*Alev*: For instance, a video can be shown to students. Students can learn like that, they can do experiments.

The lack of examples despite repeated probing by the interviewer suggests that Alev's understanding of the pedagogical aspects of aims and values was fairly limited. This is not surprising considering that her prior descriptions and drawing about aims and values were also rather limited to begin with, so it would be expected that she would find it challenging to think about the implications for teaching and learning. This observation is in contrast to her drawing and text produced after the intervention when Alev drew a boomerang to illustrate aims and values of science. In the drawing she used the words "cognitive", "epistemic" and "social" around the boomerang. She explained that she drew a boomerang to represent the relationship among three aspects of aims and values of science. She continued her explanation by stating that "all of these three aspects are interconnected. Boomerang can turn, this collect information around then turn back to science" (see Fig. 6.2).

After the intervention, when the interviewer asked Alev about aims and values of science, she said that a few points came to her mind and stated that aims and values

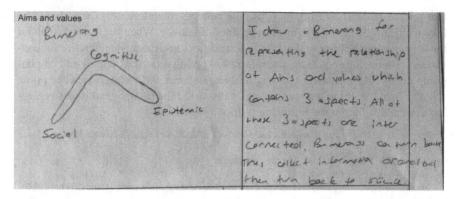

**Fig. 6.2** Alev's drawing of aims and values of science (post-intervention)

of science are interrelated concepts. She mentioned students see the effect of this information on society and stated that she thinks that this is more about values of science. Alev summarised her ideas about aims and values of science by suggesting that aims and values of science are nested and affect each other. When the interviewer asked Alev to give some examples of aims and values of science, she said that the aims of science are making science, reaching information and then transforming this information to be beneficial. She also mentioned that the aims of science affect the values of science.

> *Interviewer*: What comes to your mind when I say aims and values of science?
> *Alev*: When you say aims and values of science, a few points come to my mind. For example, firstly, aims and values are so interrelated concepts. First of all, students should be constructed science perception that means learning science, learning scientific information, learning how scientific knowledge constructed and reached. On the other hand, I think they see the effect of this information on society or seeing the effect on society. I think it is more about the values of science. In this way, I think aims and values of science are nested and affect each other.
> *Interviewer*: So, could you give any examples to aims and values of science? What are the aims and values of science through an example?
> *Alev*: The aim of science is firstly to make science and to reach information. Then, I think this information should transformed as a beneficial one. By this way, I mean the aims of science affects the values of science. Like this.

Hence, the theme of "interconnectedness" and "utility for humans" seems to be similar to her pre-intervention statements. Her reference to how learning science should demonstrate how scientific knowledge is constructed and reached suggests that she was now thinking about aims and values in relation to development of science. When the interviewer asked Alev, "Do you think aims and values of science are taught in science lessons?", she replied "no". Then, the interviewer asked about the way to teach aims and values of science in science lessons. Alev answered this question by stating that she believes that aims and values of science could be applied in science lessons by integrating them into lesson plans. She mentioned that if aims and values of science are taught as a separate subject, this could lead to a perception that science is different from its aims. She believes, therefore, that there should be lesson objectives referring to aims and values of science. When she was asked about the way to integrate aims and values in science lessons, she made reference to *models and visuals*. Subsequently, she referred to giving related examples and emphasising the effects of models as a way to integrate aims and values of science into a science topic. After the intervention, Alev discussed the pedagogical issues regarding aims and values in the following fashion:

> *Interviewer*: Ok. Do you think aims and values of science are taught in science lessons?
> *Alev*: No.
> *Interviewer*: You do not think that they are taught.
> *Alev*: I do not think they are taught.
> *Interviewer*: So, how can aims and values of science be taught in science lessons? What can be included in science lessons?
> *Alev*: I believe they can be applied to science lessons by integrating them in lessons plans because when it is taught as a different subject, it can lead to a perception that, science is different from its aims. I think it should not be taught like that. Instead, apart from the lesson objectives, there should be also objectives referring to aims and values of science.
> *Interviewer*: So, how can you integrate aims and values of science through an example?
> *Alev*: For example, what can I say… For instance, we use so many models and visuals in science lessons. We can make contributions by using these visuals. We can integrate aims and values issue into science topics and give related examples. Or, I think we can teach aims and values of science by emphasizing the effects of models.

In summary, from Alev's pre- and post-intervention drawings of aims and values of science, it is seen that she only focused on an example scenario (i.e. medical context) to represent aims and values of science before the intervention. However, after the intervention, she presented the analogy of a boomerang to represent cognitive, epistemic and social aspects of aims and values of science. The use of the boomerang suggests that she was interlinking the various aims and values. With respect to Alev's interview results, before the intervention, Alev mentioned (a) "facilitating and contributing to people's lives" and (b) "being worthwhile" as aims and values of science. After the intervention Alev thought that aims of science are making science, reaching information and then transforming this information to a beneficial one and the effect of this information on society. Even though Alev was able to refer to key categories of aims and values, the similarity of the statements before and after the intervention suggests that the precise nature of these examples did not shift in her understanding. For example, even though example epistemic aims such as "objectivity" and "empirical adequacy" were covered in the sessions, she did not make reference to any specific examples covered in the teacher education session on aims and values. However, she seems to have integrated the terminology around the types of aims and values as being cognitive, epistemic and social. Furthermore, she recognised the theme of interconnectedness among the various types of aims.

## 6.3   Representations and Perceptions of Scientific Practices

In her drawing before the intervention, Alev drew pills, serum and a molecule of phenol (i.e. an organic compound) to indicate her representations of scientific practices (see Fig. 6.3). The accompanying text emphasised applications of chemistry in medicine.

In the pre-intervention interview, when Alev was asked about scientific practices, she defined scientific practices as "practical knowledge derived from life into science". In other words, she had a sense of scientific practices with an emphasis on practical in the sense of hands-on investigations and examples that have everyday relevance, although she also expressed doubt about whether or not she was right.

*Interviewer*: Okay. What comes to your mind when I say scientific practices?
*Alev*: When you say scientific practices, I imagine practical knowledge derived from life into science but I may be wrong.
*Interviewer*: Okay. Can you give examples to scientific practices?
*Alev*: For example, boiling temperature of water is 100 degree Celsius and if we put salt in it the boiling point rises. Freezing point of water is 0 degree Celsius… When roads get slippery we throw salt. These are examples of scientific practices.
*Interviewer*: Okay, what are scientific practices? What can you say as a definition or key points?
*Alev*: Scientific practices, practical knowledge about science.
*Interviewer*: Practical knowledge?
*Alev*: Yes.
*Interviewer*: You mostly focus on scientific knowledge.
*Alev*: So, yes.

When the interviewer asked Alev "Do you think scientific practices are taught in science lessons?," she replied, "No, I don't think they are taught explicitly". She referred to the example about boiling and freezing points that she gave before and maintained that students don't learn these by trying and they only learn scientific knowledge without practising it. She repeated the "practice" idea by stating that "I see deficiency in turning these into practice". When asked about what could be

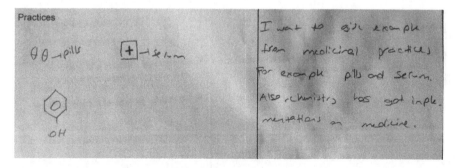

**Fig. 6.3** Alev's drawing of scientific practices (pre-intervention)

included in lessons to compensate for those deficiencies, she proposed to do more experiments and to use visual tools like videos and films.

*Interviewer*: Do you think scientific practices are taught in science lessons?
*Alev*: No, I don't think they are taught explicitly.
*Interviewer*: So scientific practices can be taught in science lessons, what do you say?
*Alev*: Let's talk from chemistry perspective. From the example I just gave, we don't learn these by trying. Only, we learn that if another substance is put in it, boiling point of the water rises while the freezing point decreases. But I see deficiency in turning these into practice.
*Interviewer*: Okay, what do you think could be added to the lessons to compensate for those deficiencies?
*Alev*: Experiments for one. I think they are crucial and we don't use them enough. As I said some visual tools may be used like videos and films. Children can watch movies.

When we turn to Alev's data from post-intervention, we witness a shift in her understanding. After the intervention, Alev drew a hexagram to represent scientific practices and wrote a scientific practice (i.e. explanation, real world, activities, data, model, prediction) at each point and "representing, reasoning, discourse, social certification" in the centre of the star. This representation is an adaptation of the Benzene Ring Heuristic that was introduced in the scientific practices session. In her explanation of her drawing, she stated that she drew a star with six ends to represent relationship among practices and all of them are important (see Fig. 6.4).

After the intervention, the interviewer asked Alev about scientific practices, and she explained scientific practices as the steps that lead to scientific knowledge. Then she referred to knowledge, data, real world, observation, testing and modelling as scientific practices. When Alev was asked to give examples for scientific practices, she again mentioned some scientific practices such as observation, and she stated that "there is a schema in my mind, other than exemplifying." She referred to Torricelli's experiment as an example from chemistry to explain scientific practices:

*Interviewer*: Okay. What comes to your mind when I say scientific practices?
*Alev*: When I say scientific practices, I think of the steps that lead to scientific knowledge.
*Interviewer*: Okay. What do you think scientific practices are?

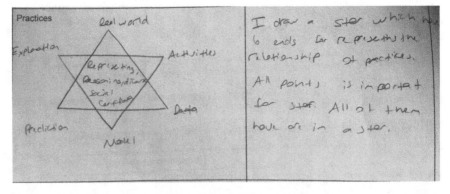

**Fig. 6.4** Alev's drawing of scientific practices (post-intervention)

*Alev*: For example, it can be knowledge, data, real world with respect to data… Was it observation? Observation. I am trying to think on scientific practices. The real world is observed and knowledge is attained. Then, there was a procedure of testing that knowledge. There can be modelling of this knowledge. There was something else but I can't remember at the moment.
*Interviewer*: Okay. Can you give examples for scientific practices? Can you explain with examples?
*Alev*: For example, as I said before, observation is made in the real world. Then, knowledge, real world, observation, gathering information. Then transfer of this knowledge, modelling…Then there is a schema in my mind, other than exemplifying.
*Interviewer*: What examples come to your mind?
*Alev*: It could be the thing from chemistry. Maybe we can mention Torricelli experiment. Torricelli makes observations while doing experiment. He looks to the real world. There is a pressure difference, then the scientific practice is this. I think it could be explained through an experiment.

In the episode above, Alev is trying to recall the components of scientific practices as described in the session, and she cannot remember all. However the fact that she is bringing in an actual example from chemistry suggests that she has made some links about the concepts related to scientific practices. Furthermore, Alev did not think that scientific practices are taught in science lessons. When she was asked about the way to teach scientific practices in science lessons, she said that she thought some topics are more suitable than others:

*Interviewer*: Okay. Okay, do you think scientific practices are being taught in science lessons?
*Alev*: No, they are not. I don't think they are being taught.
Interviewer: Okay. How could they be taught in science lessons, what could be included in lessons?
*Alev*: For example, I think some topics are more suitable to scientific practices subject. The Torricelli experiment I mentioned. While teaching gases, talking about Torricelli experiment, scientific practices can also be mentioned. I think they can be taught by incorporating into the lesson.

The comparison of Alev's drawings of scientific practices before and after the intervention indicates that the Benzene Ring Heuristic covered during the session influenced her thinking. She referred to and represented each scientific practice in a similar fashion as introduced in the session. Considering that 7 weeks had elapsed between the scientific practices session and the day of the interview, it is understandable that she could not recall all the particular aspects of scientific practices verbally but she was able to construct a visual representation that captured all the key concepts. Although Alev initially perceived scientific practices as practical knowledge with a generic reference to practice, her ideas after the intervention had a direct link to the content of the session. With respect to her perception about teaching scientific practices in science lessons, Alev was fairly limited in what she could offer as potential pedagogical strategies. She did suggest the use of experiments as a way to teach scientific practices in lessons both before and after the intervention. Even though she referred to all scientific practices to explain scientific practices, in the post-interview she only mentioned experiments as a way to teach scientific practices and not, for instance, classification which was also covered in the session. This suggests that the view of science as an experimental method persists in her perception. Moreover, given her limited pedagogical experiences, she could not provide more detail about how scientific practices could be taught to students.

## 6.4    Representations and Perceptions of Scientific Methods

When asked to draw a picture to represent scientific methods before the intervention, Alev drew pictures of test tubes, a flask and a microscope and stated that she thinks that "science can improve by making experiments. Therefore I drew experiment tubes and microscope" (see Fig. 6.5).

Hence, she used some iconic images of chemistry to communicate methods in science. In her pre-intervention interview, Alev referred to "different methods" as scientific methods. When she was asked to give some examples, she could not give any examples. Then when the interviewer probed further about chemistry examples, Alev started to talk about experiments. As an example of scientific method, she made reference to a report which is written after an experiment is conducted:

> Interviewer: What comes to your mind when I say scientific methods?
> Alev: Different methods.
> Interviewer: Could you say more? You could give some examples?
> Alev: I can't think of any.
> Interviewer: You could think of chemistry examples. What methods are used in chemistry?
> Alev: For instance, in chemistry you can do an experiment. You can produce a report after the experiment. You can write a report which is a method. But in education there's usually a statement which is later tested to see if it fits. If it does, then the experiment continues. If not, then the statement is revised. These could be scientific methods.
> Interviewer: Could you say a bit more about what methods do you think are used in science?
> Alev: Like I just said, there are methods. I am sure there are but I don't know.

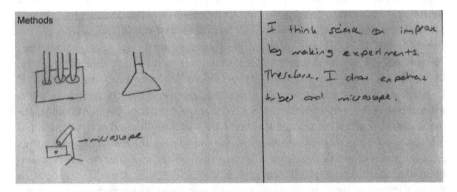

**Fig. 6.5**  Alev's drawing of methods of science (pre-intervention)

The idea of testing in the educational context was introduced but was not followed up. Alev seems to think that there should be different methods but she was not entirely clear what and why. When the interviewer asked Alev about teaching of scientific methods, she was rather firm about the fact that they are not taught in lessons but that students could be encouraged to do experiments and write up reports as an introduction to the scientific method:

> *Interviewer*: Okay. Do you think scientific methods are taught in science lessons?
> *Alev*: No.
> Interviewer: They are not taught. Right so how can they be taught then?
> *Alev*: Like the students could do experiments and write up reports. This could be an introduction to the scientific method.
> *Interviewer*: What could be included in lessons about this topic?
> *Alev*: Experiments could be included. It looks like they are included but they are not. There is no time. Students can do experiments in class. They can write up their reflections. This would contribute to learning the scientific method. They can do observations. They can write down their observations. They can learn the lesson, the topic and at the same time the scientific method.

Alev's reflections about the pedagogical aspects of scientific methods before the intervention suggest that she is mindful of the time constraints under which teachers work but that somehow teachers could integrate "the lesson, the topic and the scientific method". Following the intervention, her figure included a pencil case and pencils where she labelled the pencils as "manipulative, non-manipulative, hypothesis testing and non-hypothesis testing" (see Fig. 6.6). She explained that she drew a "pencil case model for representing methods". She further explains that the "researcher can choose each of them one by one. Every one has an effect on their research".

After the intervention, when the interviewer asked Alev about scientific methods, Alev said that stepwise method came to her mind and she referred to her learning experience in 9th grade. Then she emphasised that in the teacher education sessions,

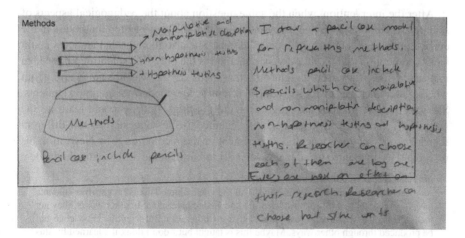

**Fig. 6.6** Alev's drawing of methods of science (post-intervention)

she learnt that going step by step is not the only method and there could be other methods:

> *Interviewer*: What comes to your mind when I say scientific methods?
> *Alev*: When we say scientific methods, the first thing that comes to my mind is the stepwise method for reaching knowledge that we learned in 9th grade. You observe, theories are constructed and then the theory is tested. Then it turns into a hypothesis. This comes to my mind. But after our workshop, I learned that in fact you don't have to go step by step and there can be other methods. In one case you could do an experiment progressing in a stepwise fashion. In another case, you could have a thesis before an observation and then you make an observation and see if you can observe it. You can follow other steps. This comes to my mind.
> *Interviewer*: Okay. Good.
> *Alev*: I learned this in the workshop.

The above quotation suggests that Alev is being reflective about her own learning of the scientific method as a linear, step-by-step process but that the teacher education session helped her think about other types of methods. She explicitly attributes her learning to the content of the session. Given that the diversity of scientific methods was a key emphasis in the session, her reference is not entirely surprising. Her use of the words "manipulative" and "modelling" is encouraging, again, considering the amount of time that had elapsed between the session and the date when the interview was conducted and the drawing was made. Furthermore, she was able to articulate further how the processes of testing and modelling might be done in different stages:

> *Interviewer*: Could you give some examples of what methods are used in science?
> *Alev*: What methods are used in science? Manipulative...do you mean that?
> *Interviewer*: Whatever comes to your mind.
> *Alev*: For example, you could do modeling. You could build up a system and then look at the variables. These are different things. You may not use models. They can happen at different times. It's not as if they have to observe first and then determine a problem and figure out the variables to test. It doesn't have to be like that. It can progress in a different way.

After the intervention, when the same question about the pedagogical aspects of scientific methods was raised, Alev indicated that the step-by-step method is in the curriculum and it is taught in science lessons. She also thought that students could have a perception that the step-by-step method is the only way scientific knowledge is reached since they were taught only that method. She was not quite sure about whether the other ways which produced scientific knowledge are taught to students. She said that even if they are taught, she did not think students could learn it. She gave some examples to teach scientific methods. She suggested to tell different scenarios such as the Torricelli's experiment or the method used by Newton in his research. Here she introduced a new scientist to build on the different methods idea:

> *Interviewer*: Do you think scientific methods are used in science lessons?
> *Alev*: Actually the step-by-step method is in the curriculum. This is taught. But students have a perception that this is how scientific knowledge is reached. In other words, they get the impression as if you can't reach knowledge any other way. But scientific knowledge can be produced through other ways. Maybe this is taught but I don't think it's learned by students in this way.

*Interviewer*: Okay. So how then could we teach scientific methods in science lessons?
*Alev*: This topic is taught as if it's a separate thing. In other words, there is such a topic and this is how scientific knowledge is produced. But it doesn't have to be that way. In a lesson plan or in a lesson, scientific methods can be included and the students themselves could do experiments or do observations and see the relationships themselves.
*Interviewer*: For example what would you include in such a lesson? What would you do?
*Alev*: For instance, I could give the example of Torricelli experiment. What he did. Or Newton and how the apple fell on his head. His method could be discussed. We could use these scenarios to talk about scientific methods.

Alev's drawings and interviews suggest that she shifted in her understanding of scientific methods. While her initial depictions were iconic representations like test tubes and flasks, her subsequent analogy of the pencil case and the diversity of scientific methods suggest that the teacher education session was getting her to use a more symbolic representations to communicate the "fit for purpose" idea about scientific methods. In other words, scientific methods are not only diverse (i.e. there are different pencils), but they are used for different purposes. She explicitly reflected on her own learning and articulated how the diversity of methods in science is not taught in lessons. While Alev perceived scientific methods as experiment and hypothesis in her pre-interview, after the teacher education intervention, she was able to provide more nuance by making reference to models and different stages of testing and modelling as well as observations. In terms of the shifts in her thinking about the teaching aspects of scientific methods, the data point to her initial perceptions as being primarily about students doing experiments and writing reports, whereas she shifted to a view of giving students a chance to do their own investigations and for lessons to integrate methods into topics. She was keen to represent the different methods through the work of different scientists and suggested the inclusion of these scenarios in teaching.

## 6.5 Representations and Perceptions of Scientific Knowledge

Similar to her drawings and verbal descriptions in relation to aims and values and scientific practices, Alev used a medicine example in visualising scientific knowledge. She drew DNA and virus models and labelled them (see Fig. 6.7). She further

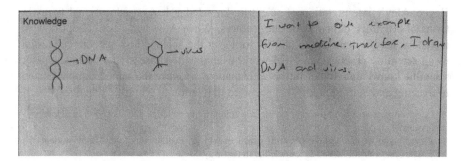

**Fig. 6.7** Alev's drawing of scientific knowledge (pre-intervention)

explained that she wanted "to give an example from medicine" to justify her drawing.

In her pre-interview, when the interviewer asked Alev, "What comes to your mind when I say scientific knowledge?", Alev replied, "science related and proven knowledge". When she was asked about the forms of scientific knowledge, she could not give an answer. Then the interviewer asked her to think about some examples. She referred to levels of knowledge that could be understood by everyday people and experts:

> *Interviewer*: What comes to your mind when I say scientific knowledge?
> *Alev*: When you say scientific knowledge, science related and proven knowledge come to my mind.
> *Interviewer*: What are the forms of scientific knowledge? Can we separate them? What would you say?
> *Alev*: Hmm...
> *Interviewer*: Maybe you can also think about some examples. Can you consider the scientific knowledge and its forms?
> *Alev*: Maybe I can think as a simple level knowledge and a high level knowledge. Simple level knowledge can be learnt or perceived by everybody. The higher level knowledge is a kind of knowledge that is perceived and produced by experts and specialist.
> *Interviewer*: So you are saying that it's about the level of difficulty.
> *Alev*: Yes.

When the interviewer asked Alev whether scientific knowledge is taught in science lessons or not, she said, "I think the simple knowledge is taught, I mean the basic level of the forms of the knowledge is taught". So the pedagogical aspects of her articulation of scientific knowledge were also based on the same categorisation as the basis of levels of knowledge. She explained her idea by linking to scientific knowledge in relation to its level of difficulty and suggested that students learn the simple level scientific knowledge but they are not aware of what they learn:

> *Interviewer*: Do you think that scientific knowledge is taught in science lessons?
> *Alev*: I think the simple knowledge is taught, I mean the basic level of the forms of the knowledge is taught.
> *Interviewer*: In your opinion, how can we teach scientific knowledge in science lessons?
> *Alev*: I think it is taught, but students don't realize that.
> *Interviewer*: What do you mean?
> *Alev*: I mean, I just divide the scientific knowledge in relation to its level of difficulty. I think the scientific knowledge that students learn is the simple level scientific knowledge. They are learning these kind of knowledge, but they aren't aware of that.

Hence, Alev's interpretation of scientific knowledge had an educational focus in terms of cognitive demands on the learners. When Alev was asked about how to include scientific knowledge and its forms in the science lessons, she suggested using scientific practices and real-world examples to make students conscious of scientific knowledge. In her further explanation, she gave the same example related to freezing and boiling points that she used in relation to scientific practices.

> *Interviewer*: I see. So, how can we include scientific knowledge and its forms in science lessons?
> *Alev*: Maybe we can add the scientific practices which I explained before. I think students can be conscious of scientific knowledge in that way.

*Interviewer*: What do you mean? Can you explain with an example?
*Alev*: For instance, we can say that this is scientific knowledge and this comes from our daily life... For instance, we can consider the boiling point of the water. If we put a foreign substance into the water, the boiling point will be increased. Just like that. Using real world examples and illustrating the connection of the topic.

Here, again, Alev is considering scientific knowledge in an educational sense, as scientific information that is made relevant to the students. After the intervention, Alev's representation of scientific knowledge consisted of a "bird bath" illustrating the flow of water across different levels (see Fig. 6.8). The figure made reference to scientific knowledge in terms of *TLM* (the terminology introduced in the scientific knowledge session), and Alev specifically linked the drawing to what "we did in class". In her explanation, she referred to models, laws and theories as scientific knowledge and the relations between them. She further gave the example of the atomic models and used the words "bigger and bigger" to communicate growth of knowledge.

In her post-interview, when Alev was asked about scientific knowledge, she said that "data and knowledge that are acquired by using scientific methods". The next question was related to the forms of scientific knowledge. Alev replied this question by stating that she did not remember. Then, the interviewer suggested to her to think about the course, and Alev began to remember about the forms of scientific knowledge. She made reference to the visual tool on scientific knowledge used during the session (see Chap. 2) and talked about theories, laws and models. She referred to the group's final project where these forms of knowledge were interlinked in an analogy:

*Interviewer*: What comes to your mind when I say scientific knowledge?
*Alev*: When I say scientific knowledge, it comes to my mind the data and knowledge that are acquired by using scientific methods.
*Interviewer*: What are the forms of scientific knowledge?
*Alev*: The forms of scientific knowledge... I don't remember right now.
Interviewer: Maybe you can think about the course to remember?

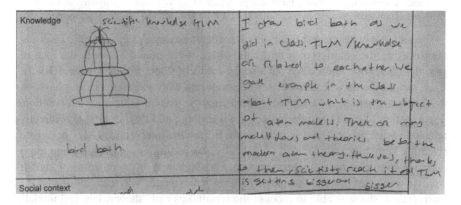

**Fig. 6.8** Alev's drawing of scientific knowledge (post-intervention)

*Alev*: Scientific knowledge...Just a second... There were nested things. There were a growing image and a bird pool from our final project. Scientific knowledge is composed of theory, law and models. They are interrelated to each other and they create the scientific knowledge.

Hence, Alev retained the theme of interconnectedness of theories, laws and models, and she seems to understand that together they form scientific knowledge. When the interviewer asked her, "Do you think that these forms of scientific knowledge are taught in science lessons?", she replied, "It is mentioned, but I think it isn't taught completely. It is not mentioned the processes and the progression of it, but there is a superficial focus." Then the interviewer asked Alev about the way to teach scientific knowledge in science lessons, and she referred to growth of scientific knowledge by giving an example from chemistry. In her explanation, she mentioned models, theories and laws about the atom to explain how scientific knowledge about the atom is constructed. She also emphasised the importance of teachers' awareness about scientific knowledge:

> *Interviewer*: OK. Do you think that these forms of scientific knowledge are taught in science lessons?
> *Alev*: It is mentioned, but I think it isn't taught completely. The processes and the progression of scientific knowledge aren't mentioned. Rather, there is a superficial focus.
> *Interviewer*: How can scientific knowledge be taught in science lessons? How can we include it in science lessons?
> *Alev*: Actually, scientific knowledge is the most convenient aspect in terms of including it into the science lessons. For instance, there are atom models in chemistry. When we teach about atom models, we can mention the related theory and law. Students generally have a view that one atom theory has gone and another one has come. However, actually one theory is getting bigger and created another one. Unless we have the previous theory, we couldn't have the other theory too. I think that students should know that. So, teaching or mentioning this is pretty easy, but teacher should be conscious about this issue. If they are aware of it, they can do it.
> *Interviewer*: OK. Can you give more examples? How can we teach scientific knowledge in science lessons? How could you integrate it into the lesson?
> *Alev*: For instance, we can consider the atom models topic as I said before. For such topic, it is important to mention about the development of models, theories and laws to teach scientific knowledge. I think, these are processes of growth, not step-by-step order of things. They are connected to each other and they are growing and progressing together.

Overall, Alev progressed from giving particular examples of scientific information to using the analogy of a bird bath where she captured not only the idea of growth of scientific knowledge but also the forms of scientific knowledge. Initially, Alev viewed the task in terms of science concepts with a subsequent emphasis on the epistemic aspects of knowledge following the teacher education intervention. The adoption of the TLM terminology from Erduran and Dagher (2014) is directly related to what was covered in the session. Before the intervention, Alev perceived scientific knowledge as the level of difficulty, suggesting a cognitive demand interpretation from an educational perspective. She explained different forms of knowledge in terms of simple versus high-level knowledge, whereas following the intervention, she was able to recount the terminology of "theory, law and model" and indicate that they are different forms of scientific knowledge. Regarding the inclusion of scientific knowledge in science lessons, Alev thought that only simple knowledge is taught to students in lessons before the intervention and students were not aware of what they learnt. She suggested using scientific practices and real-world examples as a way to include scientific knowledge in science lessons. After the intervention, she stated that scientific knowledge is taught in science lessons but not in a complete way. She thought that the processes and progression of knowledge are not mentioned in lessons, and she suggested teaching theories, laws and models by linking them. She also mentioned the necessity of emphasising growth of knowledge.

## 6.6 Conclusions

The chapter focused on one pre-service chemistry teacher, Alev, who participated in the teacher education intervention described in Chap. 4. She was part of the group of four pre-service chemistry teachers whose data were presented through a thematic analysis in Chap. 5. Alev's perceptions and representations suggest that although she was the least academically able participant in the group in terms of her GPA, she was able to appropriate some fairly sophisticated epistemic themes. She was able to link these ideas to her own examples and suggest pedagogical strategies for their incorporation in lessons. Her data contribute to our understanding of how the particular content of the teacher education sessions (e.g. diversity of methods, growth of knowledge) was interpreted and the impact of the sessions on Alev. In other words, the chapter illustrated how the instructional goals and content in the intervention had direct and close impact on Alev, illustrating a close link between the "input and output" of key themes.

The qualitative analyses presented in this chapter, as well as those in Chap. 5 provide an indication of how pre-service chemistry teachers engage in tasks around epistemic themes and how they interpret them. The data illustrate how pre-service teachers made sense of complex epistemic ideas through everyday scenarios, using analogies and visual representations to communicate their ideas. From a chemistry point of view, the data may appear limited although Alev in this chapter as well as the other pre-service teachers in Chap. 5 referred to chemistry concepts to illustrate their examples. It is worthwhile to remember that the primary objective of the intervention was to infuse some core epistemic ideas into teacher education. Hence, the emphasis was more on the development of epistemic thinking in the first place although the pre-service teachers were encouraged to use chemistry examples. Both Chaps. 5 and 6 provide empirical evidence on what is possible to accomplish in pre-service teacher education if the objective is to promote an epistemic mindset in pre-service chemistry teachers.

Further iterations of the teacher education sessions can be designed for subsequently embedding more chemistry content depending on the context and content of the programme involved. For example, once pre-service teachers grasp the notion that chemistry operates through particular epistemic aims and values, they can then be introduced to more nuanced chemistry contexts potentially through the use of historical case studies to supplement the approaches described in Chap. 4. For example, aims and values in chemistry can be unpacked through a historical case study surrounding the work of Bohr (Heilbron & Kuhn, 1969). In a study of chemistry textbook analysis, Niaz and Rodriguez (2000) discuss how many chemistry textbooks prefer to emphasise an explanation of the Balmer formula for the hydrogen line spectrum because it fits into an empiricist account of science. However two further potential accounts could also be considered: (a) quantised version of the Rutherford model of the atom and (b) explanation of the paradoxical stability of the Rutherford model of the atom. This episode contextualises epistemic aims such as empirical adequacy of models, and it also illustrates how alternative explanations can be debated.

Although such historical case studies can be used to embellish epistemic thinking, it is questionable how they can serve the needs of pre-service teachers who, like Alev in this chapter and the others in Chap. 5, have limited understanding of meta-perspectives on chemistry, including history of chemistry. Arguments for the inclusion of epistemic perspectives in chemistry teacher education (and more broadly the inclusion of philosophy of chemistry) need to take seriously the empirical research evidence on pre-service teachers' cognitive levels and background knowledge in chemistry as well as familiarity with history and philosophy of chemistry. Teacher educators, on the other hand, need to be cognizant of the baseline understanding that is demonstrated by pre-service teachers if they aim to support pre-service teachers' more nuanced and situated understanding. For example, given what we now know about Alev, subsequent interventions in her professional development can build on her newly acquired and preliminary understanding of epistemic themes to infuse progressively more sophisticated details about chemistry.

However, even if teachers can learn and internalise some abstract and unfamiliar epistemic and historical accounts of chemistry, teaching them to students in schools will be a different matter as suggested by research on the distinction between teachers' subject knowledge and their pedagogical content knowledge (see Chap. 3). The approaches described in this book can potentially help other teacher educators in developing future interventions to support their pre-service teachers in applying their understanding in teaching practice. Chemistry teacher educators, then, become a critical component of the mission of transforming perspectives from philosophy of chemistry into teachable content that serves the purposes of teacher education. In Chap. 7, we turn to a reflective account of how we, as teacher educators, have come to be involved in the teacher education intervention and what our self-study might offer to other teacher educators who might be interested in pursuing the incorporation of epistemic themes in their teaching of pre-service teachers.

# References

Elliott, J., & Lukes, D. (2008). Epistemology as ethics in research and policy: The use of case studies. *Journal of Philosophy of Education, 42*(S1), 87–119.

Erduran, S., & Dagher, Z. (2014). *Reconceptualizing the nature of science for science education: Scientific knowledge, practices and other family categories.* Dordrecht, the Netherlands: Springer.

Hamilton, L., & Corbett-Whittier, L. (2013). *Using case study in education research.* London: Sage.

Heilbron, J., & Kuhn, T. S. (1969). The genesis of the Bohr atom. *Historical Studies in the Physical Sciences, 1*, 211–290.

Miles, M. B., & Huberman, A. M. (1994). *Qualitative data analysis: An expanded source book* (2nd ed.). Thousand Oaks, CA: Sage.

Niaz, M., & Rodriguez, M. A. (2000). Teaching chemistry as rhetoric of conclusions or heuristic principles – a history and philosophy of science perspective. *Chemistry Education: Research and Practice in Europe, 1*(3), 315–322.

Yin, R. K. (2003). *Case study research: Design and methods* (3rd ed.). Thousand Oaks, CA: Sage.

# Chapter 7
# Learning and Teaching About Philosophy of Chemistry: Teacher Educators' Reflections

## 7.1 Introduction

Almost 20 years ago, Sandra Abell asked the following question and invited science educators to reflect on their career progressions: "What does/should the professional development of the science education professoriate look like? Why is it that science educators have little to say about their own and their graduate students' professional development?" (Abell, 1997). Around the same timeframe, some institutional attempts were made to establish professional standards for science teacher educators. For example, the Association of the Education of Teachers in Science (AETS) developed standards for those involved in the preparation of science teachers. These standards were meant to provide some guidelines on what science teacher educators should possess as skills, knowledge and experiences. History, philosophy and sociology of science were highlighted as a critical aspect of science teacher educators' knowledge set.

> Standard 1.d. The beginning science teacher educator should possess levels of understanding of the philosophy, sociology, and history of science exceeding that specified in the reform documents. (AETS, 1997, p. 236)

Despite such calls for the articulation of science teacher educators' practices and views, research literature in this area has been fairly limited. Irez (2006) adapted a reflection-oriented qualitative approach to study pre-service teacher educators' views of nature of science (NOS). The results of his study indicated that the participants had inadequate conceptions regarding NOS. The majority of these inadequate conceptions were concentrated under two aspects of NOS: scientific method and the tentativeness of NOS. The participants' inadequate conceptions appeared to be linked to a lack of prior reflection about NOS. Through case studies of future science teacher educators, Schwartz, Skjold, Akom, Huang, and Kagumba (2008) explored key turning points as the participants developed personal orientations towards NOS. The participants were culturally diverse, representing five different

© Springer Nature Switzerland AG 2019
S. Erduran, E. Kaya, *Transforming Teacher Education Through the Epistemic Core of Chemistry*, Science: Philosophy, History and Education,
https://doi.org/10.1007/978-3-030-15326-7_7

countries. One of the findings of their study was that motivation for learning NOS shifted from primarily external sources such as course requirements to internal factors relating to goals as a science educator. Considering science teacher educators are primarily responsible for designing and teaching the content of science teacher education programmes, their views and perceptions of NOS are central to what future teachers get to learn about NOS. Research evidence indicates that science educators with naive and eclectic NOS understanding did not perceive the teaching of NOS as a primary goal in science education (Irez, 2004).

Self-reflection and self-study can be powerful methodological approaches in how science teacher educators assess their own understanding of history and philosophy of science so that they can begin to be proactive about impacting their teaching practice. Teacher reflection is a theoretical concept that is used widely within teacher education research and practice as an important means for teachers to scrutinise and modify their teaching practices. Schön's (1983) concepts of reflection in action and reflection on action contextualise teachers' talk and learning in the action of their classrooms, constituting the sociocultural contexts of their learning. As Jay and Johnson (2002) put it, "reflective teaching entails a recognition, examination, and rumination over the implications of one's beliefs, experiences, attitudes, knowledge, and values, as well as the opportunities and constraints provided by the social conditions in which the teacher works" (p. 20). Hence, teacher educators' own reflections (e.g. Loughran, 1996) and auto-ethnographies (Saribas & Ceyhan, 2015) have been established as methodological approaches in educational research. An auto-ethnography involves self-reflection with the aim of reporting subjective experiences through a narrative (Hamilton, Smith, & Worthington, 2008). Researchers may have different purposes in developing an auto-ethnography. In an evocative auto-ethnography, the mode of storytelling is used to construct a narrative where the research process blurs the boundaries between social science and literature (Ellis & Bochner, 2000). "Analytical auto-ethnography" on the other hand involves reflexivity with a particular commitment to theoretical analysis (Anderson, 2006). As Loughran (2005) notes:

> Self-study has thus been an important vehicle for many teacher educators to find meaningful ways of researching and better understanding the complex nature of teaching and learning about teaching. (p. 5)

In this chapter, we reflect on our experiences as teacher educators and researchers as we engaged in the teacher education intervention reported in Chaps. 4, 5 and 6. A reflective account can potentially benefit other teacher educators as they take on a similar approaches in their work. Furthermore, our reflections can potentially serve as data sources that help interpret the general project of infusing epistemic aspects of chemistry in chemistry teacher education. Finally, our reflections may help ourselves in reaching realisations that might help improve our own practices as teacher educators and chemistry education researchers.

Our approach is consistent with what LaBoskey (2004) identifies as four integral aspects of self-study. The first is that self-study is improvement-aimed and that it "looks for and requires evidence of the reframed thinking and transformed practice of the research, which are derived from an evaluation of the impact of those development efforts" (p. 859). The second is the interactive nature of self-study demonstrating that "interactions with our colleagues near and far, with our students, with the educational literature, and with our own previous work ... to confirm or challenge our developing understandings" (p. 859). The third is that "self-study employs multiple, primarily qualitative methods, some that are commonly used in general educational research, and some that are innovative ... These methods provide us with opportunities to gain different, and thus more comprehensive, perspectives on the educational processes under investigation" (pp. 859–860). The fourth revolves around the need to "formalize our work and make it available to our professional community for deliberation, further testing, and judgment" (p. 860). As Loughran (2005) points out, these four aspects demonstrate an expectation that the learning from self-study will not only be informative to the individual conducting the research but also meaningful, useful and trustworthy for those drawing on such findings for their own practice.

The thought of this book came about as we were engaged in the teacher education intervention described in Chaps. 4, 5 and 6. Previously, we had been engaged in discussions about history and philosophy of chemistry in chemistry education and co-authored a paper in this area (Kaya & Erduran, 2013) as well as a paper on nature of science (Kaya & Erduran, 2016). Although in these papers, we had done some empirical work in analysing textbooks and curriculum documents, they were mainly theoretical in nature exploring how themes about history and philosophy of science can be traced through document analysis. Our theoretical discussions found a new platform when we began to adapt ideas from nature of science and philosophy of chemistry literatures for inclusion in teacher education. Throughout the implementation of the teacher education intervention as well as afterwards, we reflected on various aspects of our experiences which were recorded in a written format. This process coincided with the writing of this book as the book gave us an opportunity to reflect more formally. In this chapter, we draw on these written accounts which have now been thematically organised to provide an overview of some of the issues that our experiences have raised for us as science education researchers and teacher educators. Given the self-reflective nature of the chapter, we are using the active voice as authors in reporting on our reflections. However, whenever we are quoting excerpts from the writing of our reflections previously, we will treat these instances as data and refer to them with our respective surnames.

The organisation of the themes is based on what researchers of teacher educators have noted previously in the literature. For instance, we were mindful about considering how we got into teacher education in the first place and how our identities as teachers and researchers impact how we approach teaching unfamiliar topics such as nature of chemistry. It is widely reported that teacher educators who have taught in schools go through a dual transition in their careers (Dinkelman, Margolis, & Sikkenga, 2006). First, they transition in the institutional context, from school to higher education. Subsequently they appropriate their identities by negotiating themselves as teachers and building up their skills as researchers. As Maguire (2010) observes, such transitions can be complex and challenging. Teacher educators, often with extensive expertise and senior positions in schools, find themselves in a new institutional environment presenting challenges to their sense of professional identity. In this sense, we were mindful of capturing our own experiences and transitions to capture the nuances in our development as teacher educators and researchers on teacher education.

## 7.2  Journey to Teacher Education

Our journeys into teacher education were not entirely unexpected. We both grew up in families where the teaching profession was almost a tradition. We both had parents and grandparents who were teachers, and they were fairly instrumental for us in considering careers in education. As we embarked on the journey into the teacher education profession, we did not immediately recognise this common ground between our personal and professional backgrounds. Initially we focused on our chemistry educator identities as academics in university departments of education pursuing similar professional interests. However, increasingly it became evident that our take on teaching and teacher education, what teachers can or cannot do, was very much shaped not only by our exposure to research but also our values stemming from our upbringing and having been surrounded by teachers:

> I grew up in a family which includes educators. My mother and my grandfather were primary school teachers. Before the current teacher education system in Turkey, there were different routes to teacher education. For example, my grandfather, Kemal Ozkan graduated from "Koy Enstitusu" (literally translates as Village Institute, a state vocational training institute for students aged 12–17, now obsolete). My mother graduated from "Egitim Enstitusu" (literally translates as Education Institute, a 2-year postgraduate teacher training college, now obsolete) and became a teacher. My father also intended to become a teacher, even registered for a teacher training programme but his circumstances did not allow him to continue. My sister graduated with a degree in vocational teacher training. As you can imagine, I have thus been inspired by my family to become a teacher. One of my most memorable encounters about teaching was in primary school. I remember when our teacher was off sick I took it upon myself, as the student leader in my class, to plan a lesson and teach it! There was no supply teacher and our class teacher was absent for a few days. I had a very strong sense of responsibility as the class leader to ensure that my classmates would not be disadvantaged because of the teacher's absence. So I consulted my mother who was a teacher in my school at the time. She showed me her lesson plans and helped me come up with some ideas. I then took charge of the class and taught these ideas to my classmates. Hence, my mother in particular was as a major role model, so much so that I knew I wanted to be a teacher after middle school, and I chose to pursue education as a career in my teenage years. I went to "Anadolu Ogretmen Lisesi" (literally translates as Anatolian Teacher High School, for students aged 14–18, still in operation though teacher training provision has since changed) and I graduated from this school. Then I graduated from a chemistry education programme. However without having any experience as a teacher, I went on to pursue a doctoral degree. (Kaya)

> Teaching was always a central part of our family life as I was growing up. My father was an English teacher and my grandfather was a primary school teacher in a small village in Cyprus. I don't really know how my grandfather, Hasan Nihat became a teacher but judging

*from the impact he has had in the community, I think he must have acquired some robust pedagogical skills. For example, it has always amazed me that although he passed away in 1948 in a village in what is now southern part of Cyprus, his fellow villagers rallied around to name a street after him in northern Cyprus where they relocated after the division of the island in 1974. A proposal to name a street after him was submitted to the municipality and the name was initiated in early 2000s so almost 50 years after he passed away. To me, this is a remarkable indication of impact as a teacher and it has always inspired me as an educator. My father, on the other hand, went to an ethnically mixed (Greek Cypriot and Turkish Cypriot) teacher training college in the 1950s after which he became a primary school teacher. He advanced in his career by moving onto secondary English teaching after he attended a diploma program at Cardiff University. He subsequently engaged in various roles as an inspector and the director of in-service teacher education. I grew up being very aware of the various challenges and opportunities that all of these various roles in teaching and teacher education my grandfather and father faced. Also, even though my mother was not a teacher, she had a natural ability as a communicator. She was a great story-teller, and I learned a lot from her about engaging an audience. So when I took up secondary school teaching myself, I felt like I knew what to do. I didn't of course. It wasn't until much later in my career when I realised how I was replicating what I had been indoctrinated into through my own teachers' teaching.* (Erduran)

Apart from inspirational family stories, we were also fairly familiar with the broader context of teacher education and teaching as an academic subject. Kaya's mother was trained as a teacher in eastern Turkey during a phase when there was national political turmoil. It was also fairly unusual for a woman to be a teacher in her generation. Similarly, Erduran's father as well as her grandfather, as ethnic Turkish Cypriots, were exceptional in having teaching jobs given the conflict at the time on the island. During the times of our family's teaching experiences, teaching was regarded very highly as a profession even though the national contexts were very different. In time, we progressed into academic roles ourselves which improved our understanding of the institutional dimensions of teacher education and raised our awareness of the national provisions surrounding the accreditation of teacher education programmes and qualification of teachers.

*I was a research associate when I had my first exposure to teacher education. This was almost 20 years ago. I have since been engaged in pre-service and in-service teacher education at 4 universities where I have been employed. I realised in time the diversity of teacher education provision not only within the same country but also internationally. For example, the 3 English universities where I have had experiences differ in how they organise their programmes, and they are also very different from the Irish system. The international experience made me realise the importance of local assumptions about what is quality teacher education.* (Erduran)

*I have been doing teacher education for 14 years. I have experiences as a research/teaching assistant in a chemistry teacher education programme in 2 universities in Turkey for 10 years and as a science teacher educator in another university where I have been working for 4 years. These three universities differ in how they deal with chemistry courses for pre-service teachers. In two universities, pre-service teachers take course in chemistry departments whereas in the third, chemistry courses are delivered in the education department.* (Kaya)

It is worthwhile noting that while we both have a background in chemistry and education, we both got into teacher education without any explicit training in teacher education ourselves. Kaya had no secondary teaching experience, and Erduran had only 2 years of experience as a teacher of chemistry and science in a high school. However, we both had teaching experiences at university level before engaging in teacher education. Apart from a range of science education courses, these experiences included being a teaching assistant in laboratory courses for pre-service teachers (in the case of Kaya) and Philosophy of Social Sciences to doctoral students in education (in the case of Erduran).

Our reflections on our own progression in teacher education coupled with our reading on the topic have made us recognise that our experiences are not unique in the sense of career transitions. Kitchen (2005) outlines his growth in understanding of teaching about teaching by learning through an extensive self-study of his transition from student teacher to school teacher, to teaching associate and finally to teacher educator. Murray and Male (2005) observe that beginning teacher educators are often struck by the change in the demands of teaching from school to university. They experience unanticipated changes, and the demands of their role may initially cause them to struggle with the expectations of university culture. However, we both got instructor roles in teacher education programmes after having received PhD degrees in science education and having conducted independent research projects along the way. While this model of transition from a research degree to an academic programme of teacher education is prevalent in some systems (e.g. the USA), it is not typical in others (e.g. the UK). In this sense, our experiences of managing research and teaching responsibilities were probably unlike many teacher educators in some systems where they have to make transitions from being teachers to being teacher educators and picking up research subsequently (e.g. Dinkelman et al., 2006). Based on her research in the UK, Maguire (2010) highlighted the difficulty of teacher educators performing some traditional academic roles, such as research. In the UK, many teacher trainers are recruited into teacher education as practising teachers, and they may not have advanced research degrees including a Master's degree. Numerous proposals have been considered in supporting teacher educators in their adaptation to the research culture of universities, including within-team support (Murray, 2008), self-study (Kitchen, 2005) and development portfolios (Koster, Dengerink, Korthagen, & Lunenberg, 2008). In our case, it could be said that a more apparent challenge in our collaboration was the relative difference in our experiences as teacher educators. Even though Kaya had some experience in teacher education, engaging in a teacher education intervention with a relatively new content on nature of chemistry required mentorship by Erduran who had more experience in the theoretical background on the subject (e.g. Erduran, 2001).

## 7.3  Background in History and Philosophy of Science

As we embarked on the journey of integrating epistemological perspectives in teacher education, we were quite aware that, as science educators, we are fairly limited in our knowledge of philosophy of chemistry. However, we also recognised that we are not unusual in our profession. Many science education researchers have fairly limited background in history and philosophy of science. We return to the wise words of Michael Matthews who has been a strong advocate of this issue for many decades:

> Teacher education, and specifically the discipline of science education, is not in good philosophical health. Despite all of the concerns and arguments that have long been known and that have been documented in this book, competence in philosophy and more specifically HPS is rare in Schools of Education, nor is their attainment much encouraged. In 1989 only four of fifty-five institutions providing science teacher training in Australia offered any HPS- related course. In 1990, of the fifteen leading centres of science teacher training in the US, only half required a course in philosophy of science; the proportion in the remaining hundreds of centres was far lower (Loving 1991). The situation in the rest of the world is no more encouraging. Thus a teacher's grasp of HPS is largely picked up in their own science courses; and it is seldom consciously examined or refined. This is epistemology by osmosis, and is less than desirable for the formation of something so influential in teaching practice, and so important for professional development. (Matthews, 2014, p. 423)

Between us, Kaya is a typical example of a science education researcher who would not have had any exposure to history and philosophy of science as part of her doctoral research process:

> I had no sufficient knowledge about nature of chemistry before. I started to learn about these issues through a writing project for the first time. I was a visiting doctoral student with Professor Erduran when I first got introduced to these ideas when she was organising a seminar series on philosophy, chemistry and education in Bristol. After a year, I started to conduct research on nature of chemistry as a PhD student. The research study was about epistemological perspectives on chemistry in chemical education and published in a special issue of Science & Education. In this research, I began to read and learn some philosophical ideas related to chemistry for the first time. I never enrolled in any philosophy course during my education. I didn't know that I could or I was never encouraged to do so. So, these issues were too new but very interesting for me. For example, I did not think about concept duality before. Or I did not think how chemical knowledge is constructed. Through this research, I got into the field of philosophy of chemistry and nature of chemistry, and started to think about how we can teach nature of chemistry to students and pre-service teachers. (Kaya)

Erduran, on the other hand, was first exposed to history and philosophy of science through her own mentor, Richard Duschl, at the University of Pittsburgh where there was an internationally recognised research centre in this area. Duschl and Erduran have subsequently highlighted the importance of the infusion of epistemological themes in science education (e.g. Duschl & Erduran, 1996; Erduran & Duschl, 2004).

> Mentorship of Richard Duschl was very influential for me in picking up history and philosophy of science. His book on "Restructuring science education: The Importance of Theories" opened up a new way of thinking for me as an educator. As my supervisor for my PhD, he encouraged me to read on the subject. Before he steered me to consider taking classes on this subject in Pittsburgh, I had never even considered that there would be formal courses on these topics. Among the courses at Pittsburgh, one on "Darwinism and Its Critiques" was really influential on me. Even though it was a biology context and I was interested in chemistry, it was a turning point for me in considering scientific ideas from a philosophical point of view. Around the same time, I got exposed to Michael Matthews' seminal book on

*inclusion of history and philosophy of science in science education. I was fortunate enough to take a course with Leopold Klopfer who had produced some seminal work in the 1960s about integrating history of science in the science curriculum. I also became aware of a chemistry education distribution list where I randomly picked up an email about an article by Eric Scerri with a thought-provoking title: "Are chemistry and philosophy miscible?" That article was probably the beginning point for me in wanting to read more about philosophy of chemistry and consider how it could be made useful for educational purposes.* (Erduran)

Despite our various idiosyncratic exposure to history and philosophy of science themes in our development as science education researchers and teacher educators, we have come to recognise the limitations of our own knowledge about meta-perspectives on science.

*Somehow, I had known theories and laws as scientific knowledge but I did not know models as the other type of scientific knowledge before. I did not consider models on the same level with theories and laws before learning this approach.* (Kaya)

*One of the powerful ideas that we developed in the book,* Erduran and Dagher (2014) *was the diversity of scientific methods. Even though I had some knowledge of this issue previously, I had not fully appreciated how ignoring the multiplicity of methods in science could eventually lead to a lack of understanding what makes science 'science'. For example, I was aware from my own experiences as a teacher and a teacher educator that students in schools as well as in universities think of biology as a wishy-washy subject that is purely descriptive and does not have the same kind of rigour as physics and chemistry. I came to appreciate that the way we teach about scientific methods is partly to blame for this misconception. All sciences use different methods that sometimes are manipulative and sometimes not. Scientists sometimes test hypotheses and sometimes they simply observe and describe, and this is ok! What matters is that scientists consolidate evidence generated through different methods.* (Erduran)

As we grappled with the limitations of our knowledge in history and philosophy of science, particular themes and tools helped us orient our thinking. One of the striking features of our recent engagement in philosophical ideas on nature of science was the emphasis on a "holistic" approach. Erduran and Dagher's (2014) book on the conceptualisation of nature of science from a Family Resemblance Approach (see Chap. 2) advocated this "holistic" approach for the various aspects of nature of science as they were concerned about the fragmented nature of science in school science. In our development of the teacher education intervention, we were quite mindful of ensuring that the tasks capture this theme explicitly.

*Both nature of science and Erduran and Dagher's approach to nature of science were new for me. Actually, I started to learn nature of science through this approach. What struck me about this approach is its holistic structure and the images representing each category as well as the whole approach. All categories are related to and complemented each other. The holistic idea is embedded in each category as well as the general approach. I knew about scientific practices as an aspect of this approach and the Benzene Ring Heuristic (BRH) as the visual representing scientific practices in the previous project in which I was a researcher. However, I still had disconnected knowledge about the other categories.* (Kaya)

The notion of a holistic approach was particularly impactful on Kaya in the context of scientific methods:

*Thus, learning this holistic story was so exciting for me. In my previous education, I had learned a single scientific method that all scientists used. With this approach, I learned different methods in different fields. So, the diversity theme of this approach was another issue that struck me. I think the challenging point was to find more examples for different categories such as aims and values, scientific knowledge and methods. In addition, the social and institutional aspect of this approach was another new thing to learn for me. Even though I had thought of some issues related to political power structures such as gender, race, and so on previously, I did not think about political power structures or professional activities as some of the aspects that affect the way scientists do science. When I started to think of linking all social and institutional aspects of science with cognitive and epistemic aspects, it became a holistic story about science and it became more meaningful.* (Kaya)

In summary, the various books and projects about the transformation of theoretical ideas into educational practice were helpful in our understanding and appreciation of the utility of history and philosophy of science in science education more broadly. The mentorship experiences were instrumental in our engagement in the first place. What we can draw from our own experiences is the value and importance of mentorship in compensating, at least to some extent, the deficiencies in the systematic development of ourselves as science education researchers and our meta-level understanding of science.

## 7.4   Experiences in Incorporating Nature of Chemistry in Teacher Education

Although we have been examining epistemic themes in chemistry education primarily in the context of textbook (Kaya & Erduran, 2013) and curriculum (Kaya & Erduran, 2016) analysis, we did not work with empirical teacher education data until we started a project with pre-service teachers on incorporating scientific practices. The project was part of a research fellowship to Erduran funded by the European Union Marie-Curie/TUBITAK Brain Circulation scheme which allowed her to carry out research at Bogazici University (Erduran et al., 2016). In this project, we developed a series of workshops to infuse the Benzene Ring Heuristic (see Chaps. 2 and 3) into the same teacher education programme (Erduran, Kaya, & Dagher, 2018). Kaya at the time was based at another university but got involved in this project. The instructor of the workshops in that project along with the assistant had subsequently reflected on their own experiences in an auto-ethnography (Saribas & Ceyhan, 2015). The strategies developed in the project were used in the professional development of teachers in Lebanon with funding from NARST (Dagher, Erduran, Kaya, & BouJaoude, 2016).

As we progressed in carrying out more empirical research in teacher education, we began to realise that we needed a shorthand reference to talk about the particular theoretical orientation to nature of science developed by Erduran and Dagher (2014). Erduran and Dagher (2014) had used the Family Resemblance Approach (FRA) to discuss how nature of science can be reconceptualised into a more inclusive and holistic account. FRA was derived from Irzik and Nola's (2014) work which in turn was influenced by Wittgenstein's family resemblance idea. The issue of differentiation of the various versions of the family resemblance framework (i.e. Wittgenstein's linguistic as well as Irzik and Nola's philosophical accounts) led us to consider how to distinguish the emphasis in philosophical versus educational/ empirical accounts. We began to refer to RFN or "Reconceptualised Family Resemblance Approach to Nature of Science" (Kaya & Erduran, 2016) which was essentially the same framework as Erduran and Dagher's (2014) work, but it acted as a useful shorthand terminology:

> *I think that saying RFN is easy rather than saying the whole name of the approach. We suggested this shorthand representation in our 2016 paper. The terminology was quite useful for my students. I supervised 4 master's theses on nature of science, in particular on students' learning of scientific practices, students' perceptions of social and institutional aspects of science, and university students' understanding of all aspects of nature of science. The three theses show how RFN can be taught in lower secondary schools. While supervising these theses, I was so excited to see the impact of the intervention on my students as well as their students. My students produced many activities and lesson materials to teach these concepts. RFN also helped my Master's students in feeling a sense of community as they engaged in their research for their theses. It's shaped our research group's identity.* (Kaya)

During the design and implementation of the teacher education intervention described in Chaps. 4, 5 and 6, Kaya experienced a firsthand need to come up with terminology to have a pragmatic impact in her teaching. Erduran, on the other hand, was dealing with the transformation of theoretical ideas developed in the book by Erduran and Dagher (2014) for use in practical educational scenarios. Philosophy of chemistry as broad domain was fairly daunting in selecting useful content in inclusion in chemistry teacher education. A focus was needed to confine the content, and Erduran and Dagher's framework provided some boundaries by focusing on four epistemic categories of aims and values, practices, methods and knowledge. Still, there was the issue of transformation of theoretical categories to the planning of the teacher education sessions. Furthermore, as mentioned previously, mentorship of a relatively more junior colleague was needed as Kaya related to the content of Erduran and Dagher's book in a different way than Erduran.

> *The transition from writing a primarily theoretical book on nature of science to designing, implementing and evaluating a teacher education project was interesting yet challenging. Even though we had provided some educational implications and applications of the philosophical ideas in our book, the actual instances of what they would look like for pre-service teachers so that they are intelligible and cognitively relevant required me to pursue teacher education research literature more closely. Furthermore, I was mindful of the fact that Ebru had limited background in the work that we had done in the book. As an author of the book, I had already internalised some of the key ideas that I thought would be worthwhile to test out in practice. Soon I realised my leadership role in the development of the programme of work.* (Erduran)

Early on in our planning of the teacher education sessions, it became evident that our respective national contexts were very different in nature but shaped our assumptions about what can be done in practice in teacher education. In England, the pre-service teacher education provision is fairly limited in terms of allowing for extra work to be trialled within the programme (see Chap. 3 for a contrast of pre-service teacher education provision in our respective institutions). The fact that we could offer an 11-week course solely dedicated to a project seemed like a fertile ground for testing new ideas.

> *In this teacher education project, I offered an elective course for pre-service science, chemistry and physics teachers. Only pre-service science and chemistry teachers enrolled in the course. In the teacher education programme where I work, the teaching programme includes many elective courses. That is why the students could enroll in this course as an elective course. I could say that the teacher education programme in my institution is flexible so you can teach any specific topic in science education. (Kaya)*

The flexibility in the inclusion and duration of courses on nature of chemistry for pre-service teachers was unlike Erduran's experience in her own institution where the pre-service teacher education programme is much shorter and it is tightly regulated against a set of government standards for teaching qualification.

> *It's amazing to me that we can design a course of this length and have a comprehensive coverage of nature of chemistry with pre-service teachers in Turkey. At my university, this is practically impossible. I have taught one session of 3 hours on nature of science within our 9-month PGCE programme which is the teacher training course. I had to do a very condensed version of some of our approaches in that session and I was not satisfied that I couldn't do more but there is nothing that can be done because this is partly due to the national regulations on the length and content of initial teacher education in England. (Erduran)*

The process of designing and implementing the teacher education intervention has thus made us recognise the nuances about how our respective national and institutional contexts differ in pre-service teacher education. We increasingly recognised that although the Turkish context had much flexibility in terms of academic input, it was much less intensive in terms of teaching practice experience of pre-service teachers.

> *Pre-service teachers can teach only 4 lessons in their internship schools. Therefore, their practice experience is very limited. I wished to observe my students while teaching nature of science or nature of chemistry issues to their students in lessons but this was not possible because of the other demands that they face in their internship schools. Because of the limited time, it is really difficult to produce a programme of work for them to teach nature of chemistry. Other objectives are preferred by their mentors as well, things like teaching chemistry concepts. (Kaya)*

Nevertheless, the fact that we could have any input into pre-service teachers' understanding of epistemic issues related to chemistry was encouraging to us even though our particular ways of framing the issues were unlike what they had experienced before. In many ways, the pre-service teachers experienced difficulties in understanding some of the content in the sessions. The scientific knowledge session in particular was very challenging to them.

> *The most challenging issue for students was the scientific knowledge category. They had difficulty in understanding the links between theories, laws and models, and to produce scientific knowledge at the beginning of the session. We asked them to list theory-law-model (TLM) sets for a specific topic in different fields like biology or chemistry. Sometimes they could find two forms of knowledge such as theory-law, or law-model but they had some problems in finding the third type of the knowledge for that topic. (Kaya)*

We observed in this session that pre-service teachers had very limited understanding of models. It is probably fair to say that they had limited understanding of theories and laws as well, but at least, they could give more examples of the latter. We tried to foster their understanding through the use of the visual tools as described in Chap. 4. Yet, these tools themselves initially required much unpacking and examples before they could be meaningful to the pre-service teachers.

> *Throughout the course, I asked students to produce visuals of the categories of RFN, which are supposed to be different from those used in the sessions. At the very beginning of the course, drawing the categories of RFN was challenging for students. But they improved their drawings over time. They integrated their previous knowledge, used analogies and daily life examples to produce their drawings. Their drawings were exciting to see because they were very creative. (Kaya)*

Overall, our reflections on our experiences in incorporating epistemic themes in teacher education made us recognise the institutional barriers we face as teacher educators. While we were able to appropriate our needs, for example, by introducing a shorthand reference as RFN to ease our terminology, we had little flexibility in impacting what duration a course could be (in the case of Erduran) or how much teaching practice (in the case of Kaya) we could reasonably expect pre-service teachers to engage in. Teaching practice is clearly a very important aspect of learning to teach chemistry in general (see Chap. 3). If pre-service teachers had the opportunity to apply their learning from the sessions to their own teaching, we could have built some discussions in the sessions for them to review what worked well and what did not in their own and their peers' teaching, thereby supporting them to consolidate their understanding further. We envisage the work presented in this book as part of an emerging research programme on transforming teacher education through tools derived from theoretical work on philosophy of chemistry along with empirical data involving teachers and students. As such, there is a long-term research and development agenda that will require further investigations to build on the evidence and the practical strategies covered in the book. It is worthwhile to note that school-based teacher educators (i.e. mentor teachers) who supervise pre-service teachers on teaching practice would become instrumental if teaching practice element was built in. Introducing the epistemic core ideas to mentor teachers would be necessary through continuous professional development if the school-based teaching practice dimension is to be effectively supported.

## 7.5   Transforming Theoretical Frameworks into Empirical Research

One of our key research challenges surrounding the teacher education intervention was the lack of reliable instruments that can be used to assess pre-service teachers' understanding of RFN. Erduran and Dagher's (2014) account is a relatively recent theoretical framework. Some aspects such as the "holistic" theme was not captured in existing tools and instruments (e.g. Lederman, Abd-El-Khalick, Bell, & Schwartz, 2002).

*The interviews and drawings were useful in seeing what my students thought in a qualitative way. However, it was not easy to evaluate their understanding in a quantitative way. I had no instruments to determine their understanding levels of nature of science based on RFN. That is why we developed an instrument entitled "RFN Questionnaire" for pre-service teachers. My Master's students adopted this questionnaire for their students in lower secondary schools. As a researcher I have experienced this development process to construct instruments on nature of science and adopting this instrument for the lower secondary schools. I have to say that the development of the questionnaire challenged my research skills because although I had experience in quantitative methodology, the context was new for me and you worry about reliability and validity.* (Kaya)

Furthermore, a significant aspect of our teacher education intervention was the use of visual tools that originated from Erduran and Dagher's (2014) book. As reported in Chaps. 5 and 6, pre-service teachers' drawings were used as a data source in interpreting their perceptions of the epistemic core of chemistry. Since we had never used an approach to data collection and analysis based on visual representations, we were unsure of what to expect.

*I have to say that when I first thought of asking pre-service teachers to draw pictures of scientific knowledge, I was not very optimistic about how clear the approach would be. I realised right away that I was too biased in having been part of producing the Generative Images of Science in our book and for a minute, I thought what if you never thought of science in this kind of way? What could you possibly draw as a picture? I think I realised that even if you had some awareness of how science and its components work from a meta-perspective, this would be a really difficult question and a very vague one at that. Then I thought these students most likely won't have that meta-perspective anyway so we are probably not going to get anything from them. Yet, I thought it was worth a try to see what they might come up with.* (Erduran)

In the design and implementation of the teacher education intervention (Erduran & Kaya, 2018; Kaya, Erduran, Aksoz & Akgun, 2019), we capitalised on Erduran and Dagher's (2014) *Generative Images of Science*. We have illustrated how we have used these images in sessions and what outcomes we got as a result of groupwork and in terms of post-intervention drawings. Considering Erduran and Dagher's (2014) proposal was a theoretical idea that had not been empirically tested previously, the data in Chaps. 5 and 6 provide evidence on how pre-service teachers make sense of these images. We note two significant observations in particular. First of all, the data point to a possibility in terms of transformation of theoretical ideas into learning tools and learning outcomes. Prior to the implementation of teacher education sessions, we did not know how the pre-service teachers would interpret these images let alone how they might represent them from their own personal points of view. In this sense, the data provide some empirical validation of the theoretical idea through evidence of diversity of images. The diversity of images points to the generative nature of these images. In other words, they can be adopted and presented in different ways. Furthermore, there were many implicit ideas about some rather deep, abstract and complex ideas present in the theoretical images. For example, there were some concepts about *interdependence* of scientific practices and *growth* as well as *coordination* of scientific knowledge, concepts that refer to some dynamic relationships. The fact that the pre-service teachers managed to capture these themes in their own personal drawings is encouraging (see Table 7.1).

**Table 7.1** Generative images of science versus personal images of science

| Epistemic core | Generative images of science | Personal images of science |
| --- | --- | --- |
| Aims and values | | |
| Practices | | |
| Methods | | |
| Knowledge | | |

*During the intervention, we used Generative Images of Science developed by Erduran and Dagher in order to explain each aspect of the epistemic core. Whenever I introduced the image of a specific aspect of the epistemic core, I observed that my students could understand the key themes of that aspect. For instance, the Benzene Ring Heuristic was very useful for them to understand interrelatedness among scientific practices and the image of scientific knowledge was successful in representing the idea of growth of scientific knowledge. I could say that the generative images of science were very effective tools to teach the epistemic core of chemistry which is an abstract topic for pre-service teachers. In a way, images made my explanations easier to communicate.* (Kaya)

Even though Kaya taught the majority of the sessions, Erduran had participated via video-link in the scientific knowledge session. Also, Erduran was in Istanbul for the final session on *Generative Images of Science* and taught the session. As described in Chap. 4, the session brought together the various images for the pre-service teachers to reflect on their learning. We were both fairly surprised about how accessible these images seemed to pre-service teachers. They were able to draw out analogies to reconstruct them with different examples, either everyday examples or scientific ones.

*After the intervention, we asked pre-service teachers to draw each aspect of the epistemic core such as aims and values, methods, practices and knowledge. We have seen that they were quite creative in representing their understanding of epistemic core as visuals. These*

*visuals were very similar to the Generative Images of Science in terms of the key ideas that we taught them in the intervention. For example, one pre-service teacher drew a pencil case and different pencils in it to represent scientific methods. She explained that these different pencils represent different methods that scientists use to conduct their research. Her explanation included the fact that scientists select the appropriate method for their research like we select the appropriate pencil in the pencil case based on our aim. So, this pre-service teacher emphasized the idea of diversity of methods in science. Actually, my students' representations after the intervention surprised me. Their drawings included the analogies that we did not cover during the intervention at all. So I think that they understood the key themes through the Generative Images of Science, and then they created their personal images.* (Kaya)

*I remember walking around and looking at some of the drawings that the pre-service teachers had produced previously. I was immediately struck by how creative they seemed to be. They were also very keen to show me what they had produced. Afterwards when I saw the post-intervention data, I was once again surprised by how the pre-service teachers were able to pick out some rather sophisticated themes about these images, like "fit for purpose" idea from the methods session and come up with an example from everyday life to communicate this idea. After a discussion with Ebru, we decided that perhaps we can compare and contrast the pre-service teachers' images with the Generative Images of Science to see what they were like. From there, the idea of the Personal Images of Science came up.* (Erduran)

Having discussed the idea of comparing the visual tools from the sessions with the drawings produced by the pre-service teachers, we contrasted them in a table. We considered this to be a powerful idea, analogous to the literature on children's misconceptions in science versus conceptions in science:

*It reminds me of the research on scientists' ideas and students' ideas. When I had begun reading the misconceptions literature, I remember being struck by this analogy that was drawn in the research community. It reminded me of the "ontogeny recapitulates phylogeny" idea although of course just in a very limited sense, just in the sense of a parallel that can serve as a goal and tool for learning. Generative Images of Science are our own inventions after all, they didn't originate from the discipline per se. We did generate these images based on our review of the literature, in trying to capture themes from the literature in philosophy of science. So maybe there is a certain parallel that can be drawn. I would need to think further about this but I find this an interesting idea that can potentially be elaborated on and explored further through research.* (Erduran)

The process of contrasting the images made us realise that broadly speaking, we are dealing with disciplinary versus personal accounts of nature of science. We considered the terminology of *Personal Images of Science* to capture how pre-service teachers viewed and represented the epistemic core (see Table 7.1). The contrast revealed two aspects of our identities: (a) as learners of epistemic perspectives from philosophy of chemistry which are theoretical in nature and (b) educational researchers interested in making sense of empirical data, in this case empirical data of pre-service teachers' drawings. The contrast also made us think about potential future studies with other participants' (e.g. secondary students, in-service teachers) representations of nature of science. *Personal Images of Science* may offer both practitioners and researchers alike new possibilities in addressing nature of science in teaching and learning. For example, they can be used as assessment tools to evaluate the participants' understanding and to consider the subsequent stages in teaching to move the participants towards disciplinary depictions. Researchers might be interested in investigating the progression in the quality of images as the participants engage more in learning environments that use such images. The criteria for evaluation might be related to the key themes intended for each image. For instance, in the case of scientific knowledge, whether or not an image consists of the idea of *growth* and *interdependence* could potentially serve as evaluation criteria. An implicit aspect of the *Generative Images of Science* and *Personal Images of Science* is that they are both about meta-perspectives that are expressed visually. This visual dimension can be exploited for further use from pre-service to in-service teacher education. For example, in-service teachers' images could be used as reflective tools in their learning of nature of science. Likewise, their misconceptions about the nature of science could be elicited through their expressions in the form of drawings.

Apart from the methodological insights that we gathered by engaging at the crossroads of philosophy of chemistry, teacher education and educational research, we recognised the importance of the transformation of theoretical ideas for practical use and how this is a creative process that requires knowledge of how teachers teach, how learners learn and how we, as teacher educators, can potentially act as mediators not only between research and practice but between traditions of research in science education. We have come to clarify our own teaching and research interests in terms of (a) the intersections between theoretical frameworks from philosophy of chemistry and teacher education so that applications of philosophy of chemistry in teacher education are evidence-based relative to both fields and (b) making some fairly abstract perspectives manageable and comprehensible at the level of practitioners, so that our work is essentially one of approximation that will inevitably mean a variation from original proposals.

## 7.6   Conclusions

Our collaboration in the teacher education intervention had an empirical dimension, and it was built on existing research collaboration which was primarily theoretical in nature with respect to philosophy of chemistry. As we progressed in the planning and implementation of the intervention, we recognised that we are bound by institutional constraints and that we had to operate within what was feasible to accomplish. It would not have been possible to incorporate the intervention described in Chap. 3 in the pre-service teacher education programme in England where Erduran continues to teach. Indeed, the full scale of the intervention is likely to be applicable only in national contexts such as Turkey that can accomodate the length of the module (e.g. Gitomer, 2013). Hence, Kaya's context provided a venue for us to explore the possibility of incorporating epistemic themes in teacher education. Although we were able to achieve this goal, we were then restricted in the sense of observing teaching practices of the pre-service teachers since their teaching practice times are so limited (see Chap. 3 for a more elaborate discussion on the contrast between our two programmes of teacher education). However, our collaboration has yielded some useful realisations as well as concrete instructional strategies and research findings. The context also motivated us to be more reflective about our own motivations and background influences as teacher educators. We began to recognise the strengths as well as the limitations of our professional experiences in dealing with the incorporation of philosophy of chemistry in chemistry teacher education. At the end of the project, we found ourselves in agreement with Irez (2006) who pointed out that:

> ...developing practicing science teacher educators' understandings of NOS requires a collaborative work environment that is committed to professional development. For science teacher educators, such an environment can be achieved at two levels: at the institutional level and at the national level. Institutions that value collaboration and sharing ideas would provide opportunities for those whose understandings are incomplete or inadequate to become aware of the weaknesses in their practices and understandings and develop them. (p. 1140)

The primary theme emerging from our reflections as teacher educators and researchers in this chapter has been that understanding the identity of teacher educators is important to consider in interpreting the design and implementation of teacher education (e.g. Swennen, Jones, & Volman, 2010). Teacher educators' identities can be complex, as our own examples illustrate. Teacher educators in higher education often involve working in different contexts, such as schools, higher education administration and research (Ellis, Blake, McNicholl, & McNally, 2011). Hence, their identities can be complex and multifaceted. Based on a meta-analysis of research evidence, Swennen et al. (2010) identified the diversity of teacher educators' identities: "teacher educators as school teachers, teacher educators as teachers in higher education, teacher educators as researchers, and teacher educators as teachers of teachers" (2010, pp. 136–137). Each of these identities may present challenges in adopting novel pedagogical approaches and content. Furthermore, given the low status of teacher education in universities, many teacher educators often feel isolated, and they may struggle in the development of a higher education teacher identity (Murray & Kosnik, 2011). The "teacher educator as researcher" identity is particularly difficult to develop for those who enter higher education directly from a teaching background. The workload associated with the role in higher education and the lack of induction and professional development are typical constraints (Murray, 2008).

In our case, having had advanced educational research training, we consider ourselves as researchers. We are researchers of teacher education. We are also teacher educators. As researchers, we have come to recognise that our research ranges from theoretical arguments to empirical investigations. Chapters 1 and 2 are examples of the kind of work that we do in interpreting and positioning theoretical knowledge on philosophy of chemistry and teacher education, respectively. However, we are also researchers who actively transform theoretical ideas into practically useful versions, for instance, in the application of a synthesis of philosophy of chemistry and teacher education in teacher education practice, as described in Chap. 3. As teacher educators, we design and teach courses like the module described in Chap. 4. As social science researchers, we employ empirical research methodologies to make sense of data collected on pre-service teachers. The data can be of various kinds including images and verbal transactions. As researchers doing in empirical work, we engage in the development of analytical tools to be able to interpret such data, as we demonstrated in Chaps. 5 and 6. As teacher educators, we recognise the importance of reflection into our own teaching practices in the way that we encourage our pre-service teachers to engage in action research (e.g. Mitchener & Jackson, 2012). As teacher educators, we had to learn some rather abstract and deep philosophical ideas ourselves before we could teach them to pre-service teachers. Our own understanding improved through teacher education practice. We reached a point where we have become very aware of the necessity for simplicity and utility in transforming philosophical ideas for educational practice. Our family histories made us aware of our educational values and motivations as teachers. Our professional history as collaborators originated as a mentor and a doctoral student. In this process, we have also had to negotiate our professional identities now as colleagues who had differentiated expertise in relation to science education research, philosophy of chemistry and teacher education. As Sandra Abell would have encouraged, the engagement in this chapter gave us the opportunity to reflect on the teaching and research influences that shaped us into being science education professoriate.

# References

Abell, S. (1997). The professional development of science teacher educators: Is there a missing piece? *Electronic Journal of Science Education, 1*(4), 1–3.

AETS Ad Hoc Committee on Science Teacher Educator Standards, Lederman, N. G., Kuerbis, P. J., Loving, C. C., Ramey-Gassert, L., Roychoudhury, A., et al. (1997). AETS (the Association for the Education of Teachers in Science) position statement: Professional knowledge standards for science teacher educators. *Journal of Science Teacher Education, 8*(4), 233–240.

Anderson, L. (2006). Analytic auto-ethnography. *Journal of Contemporary Ethnography, 35*(4), 373–395.

Dagher, Z. R., Erduran, S., Kaya, E., & BouJaoude, S. (2016, April). *Infusing scientific practices in science teacher education in Lebanon*. Presentation at the Annual Conference of NARST: A Worldwide Association for Improving Science Teaching and Learning through Research. Baltimore, MA.

Dinkelman, T., Margolis, J., & Sikkenga, K. (2006). From teacher to teacher educator: Reframing knowledge in practice. *Studying Teacher Education, 2*(2), 119–136.

Duschl, R. A., & Erduran, S. (1996). Modeling growth of scientific knowledge. In G. Welford, J. Osborne, & P. Scott (Eds.), *Research in science education in Europe: Current issues and themes* (pp. 153–165). London: Falmer Press.

Ellis, C., & Bochner, A. (2000). Auto-ethnography, personal narrative, reflexivity. In N. Denzin & Y. Lincoln (Eds.), *Handbook of qualitative research* (2nd ed., pp. 733–768). Thousand Oaks, CA: Sage.

Ellis, V., Blake, A., McNicholl, J., & McNally, J. (2011) *The Work of Teacher Education: The final research report for the Higher Education Academy, Subject Centre for Education*. ESCalate.

Erduran, S. (2001). Philosophy of chemistry: An emerging field with implications for chemistry education. *Science & Education, 10*(6), 581–593.

Erduran, S., & Dagher, Z. (2014). *Reconceptualizing the nature of science for science education: Scientific knowledge, practices and other family categories*. Dordrecht: Springer.

Erduran, S., & Duschl, R. (2004). Interdisciplinary characterizations of models and the nature of chemical knowledge in the classroom. *Studies in Science Education, 40*, 111–144.

Erduran, S., & Kaya, E. (2018). Drawing nature of science in pre-service science teacher education: Epistemic insight through visual representations. *Research in Science Education, 48*(6), 1133–1149.

Erduran, S., Kaya, E., & Dagher, Z. R. (2018). From lists in pieces to coherent wholes: Nature of science, scientific practices, and science teacher education. In J. Yeo, T. Teo, & K. S. Tang (Eds.), *Science education research and practice in Asia-Pacific and beyond* (pp. 3–24). Singapore, Singapore: Springer.

Erduran, S., Mugaloglu, E. Z., Kaya, E., Saribas, D., Ceyhan, G. D., & Dagher, Z. R. (2016). *Learning to teach scientific practices* (CPD Resource). Limerick, Republic of Ireland: University of Limerick.

Gitomer, D. (2013). *Preparing teachers around the world* (Policy information report). Princeton, NJ: Educational Testing Service.

Hamilton, M. L., Smith, L., & Worthington, K. (2008). Fitting the methodology with the research: An exploration of narrative, self-study and auto-ethnography. *Studying Teacher Education, 4*(1), 17–28.

Irez, O. S. (2004). *Turkish preservice science teacher educators' beliefs about the nature of science and conceptualisations of science education*. Unpublished EdD thesis, The University of Nottingham, Nottingham.

Irez, O. S. (2006). Are we prepared? An assessment of preservice science teacher educators' beliefs about nature of science. *Science Teacher Education, 90*, 1113–1143.

Irzik, G., & Nola, R. (2014). New directions for nature of science research. In M. Matthews (Ed.), *International handbook of research in history, philosophy and science teaching* (pp. 999–1021). Dordrecht, the Netherlands: Springer.

Jay, J. K., & Johnson, K. L. (2002). Capturing complexity: A typology of reflective practice for teacher education. *Teaching and Teacher Education, 18*(1), 73–85.

Kaya, E., & Erduran, S. (2013). Integrating epistemological perspectives on chemistry in chemical education: The cases of concept duality, chemical language, and structural explanations. *Science & Education, 22*(7), 1741–1755.

Kaya, E., & Erduran, S. (2016). From FRA to RFN, or how the family resemblance approach can be transformed for curriculum analysis on nature of science. *Science & Education, 25*(9–10), 1115–1133. https://doi.org/10.1007/s11191-016-9861-3

Kaya, E., Erduran, S., Aksoz, B., & Akgun, S. (2019). Reconceptualised family resemblance approach to nature of science in pre-service science teacher education. *International Journal of Science Education, 41*(1), 21–47.

Kitchen, J. (2005). Looking backwards, moving forward: Understanding my narrative as a teacher educator. *Studying Teacher Education, 1*(1), 17–30.

Koster, B., Dengerink, J., Korthagen, F., & Lunenberg, M. (2008). Teacher educators working on their own professional development: Goals, activities and outcomes of a project for the professional development of teacher educators. *Teachers and Teaching, 14*(5–6), 567–587.

LaBoskey, V. K. (2004). The methodology of self-study and its theoretical underpinnings. In J. J. Loughran, M. L. Hamilton, V. K. LaBoskey, & T. Russell (Eds.), *International handbook of self-study of teaching and teacher education practices* (pp. 817–870). Dordrecht, the Netherlands: Kluwer.

Lederman, N. G., Abd-El-Khalick, F., Bell, R. L., & Schwartz, R. (2002). Views of nature of science questionnaire (VNOS): Toward valid and meaningful assessment of learners' conceptions of nature of science. *Journal of Research in Science Teaching, 39*(6), 497–521.

Loughran, J. (1996). *Developing reflective practice: Learning about teaching and learning through modelling*. London, Palmer Press.

Loughran, J. (2005). Researching teaching about teaching: Self-study of teacher education practices. *Studying Teacher Education, 1*(1), 5–16.

Maguire, M. (2010). Inside/outside the ivory tower: Teacher education in the English academy. *Teaching in Higher Education, 5*(2), 149–165.

Matthews, M. R. (1994/2014). *Science teaching. The role of history and philosophy of science.* New York: Routledge.

Mitchener, C., & Jackson, W. (2012). Learning from action research about science teacher preparation. *Journal of Science Teacher Education, 23*(1), 45–64.

Murray, J. (2008). Towards the re-articulation of the work of teacher educators in higher education institutions in England. *European Journal of Teacher Education, 31*(1), 17–34.

Murray, J., & Kosnik, C. (2011). Academic work and identities in teacher education. *Journal of Education for Teaching: International Research and Pedagogy, 37*(3), 243–246.

Murray, J., & Male, T. (2005). Becoming a teacher educator: Evidence from the field. *Teaching and Teacher Education, 21*(2), 125–142.

Saribas, D., & Ceyhan, G. D. (2015). Learning to teach scientific practices: Pedagogical decisions and reflections during a course for pre-service science teachers. *International Journal of STEM Education, 2*(7), 1–13. https://doi.org/10.1186/s40594-015-0023-y

Schön, D. A. (1983). *The reflective practitioner: How professionals think in action.* New York: Basic Books.

Schwartz, R., Skjold, H. H., Akom, G., Huang, F., & Kagumba, R. (2008). *Case studies of future science teacher educators' learning about nature of science.* Paper presented at the American Educational Research Association Annual Conference. New York. March 24–28.

Swennen, A., Jones, K., & Volman, M. (2010). Teacher educators: Their identities, sub-identities and implications for professional development. *Professional Development in Education, 36*(1–2), 131–148.

# Chapter 8
# Towards Development of Epistemic Identity in Chemistry Teacher Education

## 8.1 Introduction

The book began with a review of recent research literature on philosophy of chemistry and chemistry education. The reference to philosophy of chemistry in Chap. 1 highlighted the conceptual background and justified the relevance of epistemic perspectives for inclusion in chemistry education. The idea of the "epistemic core" provided a concise lens through which epistemic themes can be selected. The categories of scientific aims and values, practices, methods and knowledge provide an inclusive and comprehensive account which can be unpacked depending on the content and the context of the curriculum. Epistemic aims and values include such concepts as accuracy and empirical adequacy. Methods involve experimentation and observation which may involve manipulative or non-manipulative approaches. Practices concern those processes that mediate the formulation of chemistry knowledge. For example, the coordination of data to generate models and the use of predictions can be considered examples of epistemic practices. Knowledge consists of forms of knowledge as theories, models and laws. A coordinated epistemological approach was presented given the bringing together of the means (e.g. practices and methods), the reasons (i.e. aims and values) and outcomes (i.e. knowledge) of chemical inquiry. Erduran and Dagher's (2014) *Generative Images of Science* were used as visual tools to organise some key themes that can be unpacked further from the epistemic core. In Chap. 2, the epistemic core idea was related to particular chemistry concepts to illustrate how it can be considered in the context of chemistry, given the domain-general nature of the original framework in Erduran and Dagher's work.

In Chap. 3, research evidence on teacher education was reviewed, particularly with respect to how teachers learn and what epistemic beliefs they hold which influence how they learn to teach (e.g. Peters & Kitsantas, 2010). The diversity of routes to pre-service teacher education was highlighted including the challenges regarding programme-level constraints in what is possible to include in the content of

© Springer Nature Switzerland AG 2019
S. Erduran, E. Kaya, *Transforming Teacher Education Through the Epistemic Core of Chemistry*, Science: Philosophy, History and Education,
https://doi.org/10.1007/978-3-030-15326-7_8

pre-service teacher education. For example, in England a recent report by the House of Commons highlights the financial and policy developments (Foster, 2018) illustrating the political dynamics surrounding pre-service teacher education. Perhaps the main common thread in all teacher education provisions is the statutory requirement specified as the national curriculum in the subject and although there may be some extent of flexibility, teachers typically teach to the curriculum in state schools. A further aspect of the complexity of teacher education relates to the existence of different routes to pre-service teacher education in many parts of the world. In England, around 30,000 individuals sign up to qualify to teach each year through a number of routes. Although they vary in other ways, "the main distinctions between the different routes are whether they are school-centred (for example, School Direct) or higher education led, and whether the trainee pays tuition fees or receives a salary. Alongside the routes currently available, a school-led postgraduate teaching apprenticeship is set to be launched from September 2018" (Foster, 2018, p. 3). The contrast of national and institutional contexts in Chap. 3 illustrated some of the challenges and opportunities for inclusion of epistemic themes in initial teacher education.

Following the theoretical background to the inclusion of epistemic aspects of chemistry in chemistry teacher education, Chap. 4 presented a description of and rationale for the design and implementation of a pre-service teacher education project that aimed to integrate the epistemic core in chemistry pre-service teachers' learning. A teacher education module was offered as part of an undergraduate course at Bogazici University in Turkey. The content of the sessions was described including the range of pedagogical strategies employed such as group discussions and presentations. The use of visual representations and analogies was encouraged for facilitating pre-service teachers' understanding and interpretation of the epistemic core idea which is fairly abstract and unfamiliar to pre-service teachers. One session on each epistemic core category was covered by an instruction who was either of the authors of the book, and it was followed up by a second session when pre-service teachers worked in groups and developed lesson ideas based on a new chemistry topic. Chapter 5 showed the particular emphases in the intervention (e.g. diversity of methods, growth of knowledge) and the ways in which pre-service chemistry teachers have been influenced by the intervention. The chapter illustrated how the instructional content was manifested in pre-service teachers' perceptions and representations.

In Chap. 6, the case of one pre-service teacher with a low GPA was illustrated to show how she engaged in ideas about the epistemic core. The choice on this pre-service teacher was deliberate. We wanted to highlight what is possible to accomplish with pre-service teachers who have limited academic skills and therefore may find epistemic themes difficult to understand. The qualitative analyses presented in Chaps. 5 and 6 provide an indication of how pre-service chemistry teachers engage in tasks around the epistemic core and how they interpret them. The data indicated that pre-service teachers made sense of complex epistemic ideas through everyday scenarios, using analogies and visual representations to communicate their ideas. This is of course not surprising given that we promoted the use of such an approach in the teacher education sessions. What the outcome illustrates is what such strategies enable teacher educators to access in pre-service teachers' developing understandings. The content of the teacher education intervention can be considered as the first step to the realisation of more complex epistemic thinking. There are implications of the pre-service teachers' learning for the learning of their students. It could be that visual representations, analogies and argumentation can be used with students as well as potential strategies to foster epistemic thinking in students. However, such an approach would need careful consideration given students are unlike pre-service teachers in that they have very limited knowledge of science to begin with. Furthermore, by their very nature, some of the approaches used (e.g. analogies) can be problematic and would need a more nuanced approach in ensuring that students do not misinterpret the intended goals of instruction.

What are the implications of learning to teach these themes by pre-service chemistry teachers? How can in-service teachers be supported in learning such themes? What are the stages in how pre-service teachers progress in their learning as they acquire more and more sophisticated understanding of the epistemic dimensions of chemistry? For example, we have observed how pre-service chemistry teachers understand the theme of diversity of methods in chemistry. How would more experienced teachers consider this theme? It is possible that given the broader repertoire of pedagogical experiences and skills of more experienced teachers, they would be drawing on more robust examples and would be in a better position to make sense of how these themes could potentially be taught in chemistry lessons. However, given the lack of content on the epistemic aspects of chemistry in both pre- and in-service teacher education, it is likely that the learning of the epistemic core will be just as challenging for in-service teachers. Our discussion raises questions that can set a new research agenda on the development of teacher expertise in terms of epistemic thinking and understanding. There are implications not only for pre-service and in-service teachers but also for the various actors who are involved in their professional development. In the case of the pre-service teachers, school-based mentors and teacher educators will also need to be supported if they do not possess sufficient background in philosophy of chemistry (see Fig. 8.1). Likewise in-service teachers will be guided by their school leaders, typically heads of science, and professional development course coordinators (see Fig. 8.2). In both cases, there will need to be programme-level developments to produce sufficient resources including training as well as teaching and learning materials.

In Chap. 7, a self-study of ourselves as teacher educators was outlined, reflecting on our own knowledge on philosophy of chemistry and professional experiences. The underpinning assumption in this chapter has been that understanding the identity of teacher educators is critical in ensuring that teacher education can be innovative and progressive (e.g. Swennen, Jones, & Volman, 2010). Teacher educators' identities can be complex, as our own cases illustrated. Teacher educators in higher education often work in different contexts, such as schools, higher education administration and research (Ellis, Blake, McNicholl, & McNally, 2011). Hence, they may have variations in their identities which can be multifaceted. Swennen et al. (2010) identified the diversity of teacher educators' identities in the following way: "teacher educators as school teachers, teacher educators as teachers in higher education,

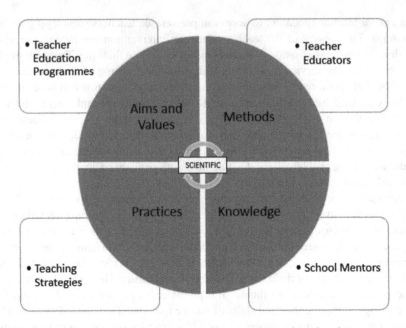

**Fig. 8.1** Systemic coverage of the epistemic core in pre-service teacher education

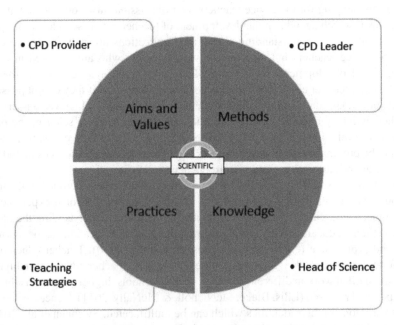

**Fig. 8.2** Systemic coverage of the epistemic core in continuous professional development of in-service teachers

teacher educators as researchers, and teacher educators as teachers of teachers" (pp. 136–137). Each of these identities may present challenges in adopting novel pedagogical approaches and content. Given the low status of teacher education in universities, many teacher educators often already experience isolation and struggle in the development of a higher education teacher identity (Murray & Kosnik, 2011), let alone take on a fairly ambitious task as ours in this project. The "teacher educator as researcher" identity is particularly difficult to develop for those who enter higher education directly from a teaching background (Murray, 2008).

Through our reflections, we have come to re-envisage ourselves as teacher educators who are trying to make sense of some rather abstract and deep philosophical ideas ourselves, pursuing both theoretical and empirical questions. In other words, we were teacher educators learning to teach about epistemic themes to our pre-service teachers. Having had advanced research training in social science methodologies, we were also positioned as researchers, in particular researchers of teacher education. Hence our identities were multifaceted as we engaged in the design and implementation of the teacher education intervention. Through our own reflections and self-study, we have come to recognise the importance of developing our own knowledge of the epistemological aspects of chemistry through exploration of the related research literature. In our reading, we were guided by particular criteria such as pedagogical relevance and cognitive demand of epistemological perspectives in chemistry on teachers and students. In other words, our selection and use of particular epistemic themes in the project were informed by our own knowledge of research in chemistry education. We were also guided by pragmatic knowledge of what is possible to accomplish in a real context of teacher education, as opposed to a theoretical, idealised or abstracted space of teacher education and teaching.

## 8.2 A Framework on Epistemic Identity

Our engagement in theoretical perspectives on philosophy of chemistry and empirical research on teacher education led to the recognition that simplicity and utility are decisive criteria in transforming philosophical ideas for educational practice. The collaboration underpinning the work had an element of mentorship. Given our own professional history was established as one of a doctoral mentor and student, we have also had to negotiate our professional identities now as colleagues who had differentiated expertise in relation to science education research, philosophy of chemistry and teacher education. Models of mentorship and collaboration among teacher educators might facilitate others as they embark on incorporating epistemic themes in their teaching. Although challenging, the task ahead of us was not unintelligible. We already have had numerous years of being exposed to the academic literature of philosophy of science. Some teacher educators may not have had such background. Furthermore, more inexperienced teacher educators are often struck by the demands of teaching as they transition from school to university (Murray & Male, 2005) leading to significant challenges presented by the university culture.

The discussion thus raises queries about identity: of teachers, teacher educators, researchers, philosophers, chemists and other stakeholders who contribute to the mission of chemistry teacher education. Hence, we now explore the discussion so far in the book within the context of the newly emerging science education research literature on identity (e.g. Carlone & Johnson, 2007; Zembal-Saul, 2016). A framework on the development of "epistemic identity" in pre-service teacher education and continuous professional development is proposed. The framework provides a comprehensive and inclusive approach to addressing the fundamental question of how future and existing teachers of chemistry can be supported in becoming inquisitive about what knowledge about chemistry is, how it is constructed and how it is justified. As the next section will stress, the science identity research has brought to the foreground the importance of considering not only teachers' knowledge and beliefs but also their identities. We as teacher educators we recognised and reported in Chap. 7 that learning, including ours, is intricately linked to aspects of one's identity. We came to recognise the significance of our own identities as teacher educators in adopting to novel content as we engaged in the empirical project of transforming theoretical ideas into practical toolkits and strategies. In other words, the process of engagement in the empirical adaptation of theoretical ideas made us recognise that identity is a significant component of the work that we are trying to accomplish. Hence, we will now consider more broadly the issue of identity in relation to the main scope of the book, which is to infuse the epistemic core into chemistry teacher education.

In considering the implications for the integration of themes from philosophy of chemistry in chemistry teacher education, research on various aspects of learning to teach, including the role epistemic beliefs, was reviewed in Chap. 3. The focus on epistemic beliefs was not entirely coincidental given the relevance of beliefs in education. Yet, epistemic beliefs and related areas of research on an assortment of related terminology such as epistemic cognition (e.g. Green, Sandoval, & Braten, 2016) tend to emphasise the cognitive, psychological and philosophical accounts of epistemological issues. It is worthwhile to consider a broader perspective on "identity" and explore the implications for teacher education. Teacher education researchers have begun to note the role of pre-service teachers' identities in the shaping of their developing expertise in teaching. In a review of the literature on identity in science education, Avraamidou (2014) states that the construct of identity is particularly important within the field of teacher education because "it offers a comprehensive construct for studying teacher learning and development, which goes beyond knowledge and skills" (p. 146). As Wenger (1998) argued, "because learning transforms who we are and what we can do, it is an experience of identity" (p. 215). Identity researchers have often built a contrast between pre-service teachers' cognition (e.g. knowledge, beliefs, understanding) and other aspects of their learning:

> The sole focus on knowledge, understanding or other purely cognitive constructs in teacher education, has been criticized as limited as it leaves the novice teacher alone to figure out how to develop, integrate, and reconcile emotions and physical aspects with the understandings involved in becoming a teacher. (Luehmann, 2007, p. 827)

Pre-service teachers need to acquire a complex set of knowledge and skills and understandings of pedagogical practices. They also need to create and recreate their image of themselves as members of a professional community (Sutherland, Howard, & Markauskaite, 2010). The subject matter can be influential in how pre-service teachers view their own identities (Helms, 1998). Volkmann and Anderson (1998) examined the year-long teaching journal of a 1st-year chemistry teacher in order to characterise the nature of her professional identity. Analysis of the participant's journal illustrated three dilemmas that she faced during her 1st year of teaching: feeling like a student while expected to be an adult, caring for students and the expectation to command respect and disliking the subject matter of chemistry while expected to be a chemistry expert. These findings showed that the participant's professional identity was connected to her history, the expectations of the school, her content knowledge and her own vision of what it means to be a teacher. It is plausible that the tensions faced by pre-service chemistry teachers in relation to the subject matter of chemistry stem from lack of preparation in terms of the "how" and "why" of chemistry knowledge, with the "what" having been dominant in their own learning as students. The sort of approach that we are promoting in this book would empower pre-service teachers so that they not only possess chemistry knowledge but also know why and how such knowledge is justified and produced.

The research literature on pre-service teachers' identities has focused on various aspects of "identity" including personality (e.g. Rodgers & Scott, 2008), professionalism (Beijaard, Meijer, & Verloop, 2004), gender and ethnicity (e.g. Bianchini, Cavazos, & Helms, 2000). Carlone and Johnson (2007) developed a science identity model which includes three interrelated and overlapping dimensions of science identity: competence, performance and recognition. Performance is used to refer to social performances of relevant scientific practices such as ways of talking. Recognition refers to recognising oneself and being recognised by others as a science person. Competence is used to refer to knowledge and understanding of science content. Carlone and Johnson used this model as an analytic lens in their ethnographic studies aiming to understand the science experiences of successful women of colour over the course of their undergraduate and graduate studies in science and into science-related careers. Another model of identity proposed by Beijaard et al. (2004) includes the following essential features: (a) identity is an ongoing process of interpretation and reinterpretation of experiences; (b) identity implies both person and context; (c) identity consists of sub-identities; and (d) agency is critical and refers to the need of teachers to be active in the process of professional development (p. 122).

Similarly, other researchers have examined identity using various theoretical frameworks including social theories of learning (e.g. Wenger, 1998), cultural historical activity theory (Roth & Tobin, 2007) and positionality (Maher & Tetreault, 2001). As Zembal-Saul (2016) argues, this range and heterogeneity of theoretical frames have contributed to multiple definitions of science teacher identity and a complicated set of overlapping constructs such as agency, authoring and positioning. While research on the role of identity in science teacher education is now fairly substantial, the particular emphasis on teachers' identities in relation to their learning about epistemic aspects of science is scarce. Akerson, Carter and Elcan (2016) contributed to this area by investigating teachers' identities as teachers in relation to the teaching of nature of science (NOS) and observed that:

> ...preservice teachers who are beginning to conceptualize NOS do not generally exhibit an ongoing process of interpretation and reinterpretation of NOS identity, nor do sub-identities necessarily infringe upon their NOS identity development. However, as they begin field experience teaching there is more evidence of ongoing interpretation and reinterpretation of NOS identity. As they begin full-time student teaching, when they are responsible for teaching all subjects, they do exhibit competing sub-identities. For in-service teachers who are in full time teaching, the framework works well to describe their identity development as they are juggling competing identities as teachers of other content areas, and must have agency to continue to emphasize NOS. (p. 97)

In relation to professional development of in-service teachers, Akerson et al. (2016) note that NOS understandings have included examining NOS as inquiry, thus teachers often develop an identity of teacher as inquirer through active investigation of science content. The authors argue that through inquiry investigations, teachers are able to explore not only science practices and science content but also NOS. Having an "active and reflective stance on their own teaching practice allows in-service teachers to take a deeper look at their science instruction, and how NOS can be infused within the curriculum that they already teach" (p. 101).

Although "identity" has recently gained momentum as a research area in science education (e.g. Avraamidou, 2014) and teachers' identities in relation to NOS have been questioned (e.g. Akerson et al., 2016), the construct of "identity" has been addressed from various disciplinary perspectives in the social sciences. For example, in social identity theory, a social identity is a person's knowledge that he or she belongs to a social category or group (Hogg & Abrams, 1988). In a social group, individuals view themselves as members of the same social category. Individuals who are similar to the self are categorised with the self, and they are labelled the "in-group", while those who differ from the self are categorised as the "out-group". In early work, social identity included the emotional, evaluative and other psychological correlates of in-group classification (Turner, Hogg, Oakes, Reicher, & Wetherell, 1987, p. 20). In identity theory, on the other hand, the core of an identity is the categorisation of the self as an occupant of a role and self-incorporation of the meanings and expectations associated with that role and its performance (Thoits, 1986).

What are the implications of the construct of "identity" for the epistemic core idea covered in this book? Can we consider an aspect of identity that concerns epistemic themes? Can a construct of "epistemic identity" be useful for chemistry teacher education? Existing interdisciplinary literature includes related terminology such as the "Epistemological Identity Theory" (EIT). As Demerath (2006) explains EIT as follows:

> Epistemological Identity Theory (EIT) explains how individuals enhance their knowledge of self and the world by creating and maintaining identities. Using cognitive and affective processes previously ignored by identity theorists, this theory reconceptualizes commitment to an identity as the degree to which that identity organizes and clarifies one's experience of the world and him or herself. (p. 491)

Such take on identity from an epistemological perspective places the emphasis on individuals as knowers of themselves. In contrast, our concern about a framework on epistemic identity is about the diversity of facets that contribute to the individuals' positioning about knowledge and knowing. As reviewed in Chap. 3, teachers' epistemic beliefs can influence how they approach knowledge. Epistemic beliefs are a cognitive account. Individuals can also have affective dispositions towards knowledge and knowing. In *Epistemology and Emotions*, Brun and Kuenzle (2008) draw attention to the affective aspects of epistemology:

> *Epistemic activities can be very emotional affairs. Curiosity, doubt, hope and fear trigger everyday cognitive activities as well as academic research, which in turn are sources of surprise, frustration and joy. Less intellectual emotions may also play their part when tireless scrutinizing is driven by jealousy, or when an experiment is too disgusting to occur to any researcher.* (p. 1)

Furthermore, as previously noted, the social identity theory highlights a person's knowledge that he or she belongs to a social category or group (Hogg & Abrams, 1988). Such a social account stresses the role of social membership and agency. In application to a construct of epistemic identity, it may points to how social mechanisms may shape individuals' participation and belonging in matters of epistemic concern. A final dimension of epistemic identity to be considered is rationality. Kelly (2003) defines epistemic rationality as "...the kind of rationality which one displays when one believes propositions that are strongly supported by one's evidence and refrains from believing propositions that are improbable given one's evidence" (p. 612). Rationality in this sense is not only about cognitive accounts of individuals' thinking but also about how institutionalisation of knowledge contributes to a set of negotiated public criteria that drive the evaluation of knowledge. This aspect of epistemic identity emphasises the institutional dimensions of epistemic identity. A framework on "epistemic identity" based on four components (i.e. cognitive, affective, social and institutional) is inclusive of the various dimensions of identity that researchers have considered more broadly.

The consolidation of the theoretical backings from an array of perspectives is appropriate and needed for educational research that itself is interdisciplinary by nature (e.g. Broggy, O'Reilly, & Erduran, 2017). Epistemic identity extends the discussion on the epistemic core (see Chaps. 2, 3, 4, 5 and 6) to demonstrate how mere consideration of the epistemic core from a cognitive perspective is likely to be limited in chemistry teacher education. Although some gains can potentially be obtained through carefully designed and implemented interventions, they are likely to be limited if the broader positioning of chemistry teachers is not considered. Our own reflections of our professional development as we conducted the teacher education project (see Chap. 7) led us to recognise the relevance the issue of "identity" in relation to learning. As teacher educators, our own identities in relation to epistemic matters came to be significant in our considerations in the design and implementation of the teacher education project. The content of the sessions was informed by our understanding of the epistemic core. This itself related to our cognitive abilities and skills in relation to understanding philosophy of chemistry. As teachers of preservice teachers, we recognised our own epistemic agency (i.e. ourselves as teacher educators who are social mediators of the epistemic core) in the social context of the sessions as well as the institutional context of the teacher preparation programme. Often our engagement with philosophy of chemistry was demanding and difficult, although rewarding at the same time.

Although we did not specifically reflect on the affective and emotive dimensions of our experience formally, we recognise that the process of learning philosophy of chemistry was not entirely emotion-free. Yet, we strived to achieve a certain standard of rationality by promoting components of the epistemic core as being underpinned by commitment to institutionalised criteria (i.e. objectivity and empirical adequacy in the case of aims and values). Hence, our point of departure in recognising the significance of the notion of epistemic identity was shaped by our experiences as teacher educators and researchers of teacher education in the overall project of this book. As mediators of knowledge and knowing processes with students, teachers (both pre-service and in-service teachers) are expected to possess similar kinds of cognitive skills, affective dispositions and social agency about epistemic matters as well as commitment to institutional criteria for epistemic rationality. In the way that our own cognitive abilities with philosophy of chemistry were challenged, teachers will be challenged to acquire unfamiliar content. A broader and systemic perspective that consolidates the epistemic, cognitive, social and institutional dimensions for the enactment of the epistemic core in teacher education recognises different kinds of support that are needed to facilitate the development of teachers' skills with teaching epistemic themes.


## 8.3   Epistemic Identity and Teacher Education

Epistemic identity can be conceptualised as a myriad of beliefs and emotions about knowledge and knowing coupled with a recognition of individuals' social agency as mediators and negotiators of knowledge and knowing as well as commitment to institutionalised rationality. The construct of "epistemic identity" is encompassing. The various dimensions point to why it may be difficult to have significant shifts in teachers' epistemic thinking with repeated professional development. For example, while it may be possible to impact on teachers' understanding of the epistemic core categories, teachers may not consistently apply them because they may not have appropriated these categories as part of their broader epistemic identity. They may have emotive responses to the use of the epistemic core in their teaching or may not fully appreciate their own social agency in the enactment of the epistemic core. The general framing of the definition of epistemic identity is consistent with Carlone and Johnson's (2007) characterisation of scientific identity in terms of competence, recognition and performance. The outcomes of these competences would allow teachers to develop sub-identities as developers, evaluators and justifiers of knowledge. Teachers' development of epistemic identities can be conceived as an ongoing process that relies on numerous systemic factors that influence the process, in line with Beijaard et al.'s (2004) model of scientific identity. For instance, teacher educators can impact the process of teachers' ongoing learning from initial teacher education of pre-service teachers to continuous professional development (CPD) of in-service teachers.

The theoretical epistemic perspectives (Chaps. 1, 2 and 3), teacher education strategies (Chaps. 3 and 4), pre-service teachers' learning outcomes (Chaps. 5 and 6) as well as our awareness of teacher educators' identities (Chap. 7) have led us to consider why a broader construct of "epistemic identity" is needed in relation to teacher education. These dimensions of our argument lead us to propose a framework for pre-service teachers' epistemic identity. Teacher educators and CPD providers (e.g. university-based teacher educators and school-based teacher mentors), institutions (i.e. university teacher education programme or school-based CPD courses) and novel teaching strategies (e.g. teaching and learning tools) all contribute to the development of teachers' epistemic identities. Programme-level constraints will need to be considered, and the content will need to be adjusted to be applicable and relevant. The familiarity and competence of the school-based teacher mentors are critical because they work very closely on a day-to-day basis with pre-service teachers. University-based teacher educators typically engage in mentor development sessions where the goals and content of the university-school partnerships are discussed, developed and reviewed. Mentors can be inducted as part of such programmes to learn about epistemic themes, and they could target mutual projects to pursue with the pre-service teachers that they will be mentoring. Particular teaching and learning strategies such as argumentation, visual representations and analogies, as highlighted in Chap. 3, can be introduced as example strategies for supporting the learning of the epistemic core.

Learning of the epistemic core will be a long-term goal beyond the acquisition of some skills in pre-service teacher education. As future teachers secure teaching jobs and settle into their working lives, they are entitled to attend CPD courses offered by a range of providers. For instance, in England, the *Royal Society of Chemistry* has a large programme of CPD courses that are offered throughout the year, and there are also professional conferences such as the *Association for Science Education* national and regional conferences aimed for in-service teachers. School management and leadership could consider a programme of work for the science teachers in their schools in targeting CPD opportunities that enhance teachers' learning of epistemic themes. School-based in-service teacher training programmes could similarly aim to devise and implement a programme of work that would provide continuous support for science teachers to engage in the development of their skills in teaching epistemic themes.

Sustained long-term and systemic engagement in professional development targeting learning of epistemic themes is likely to contribute to the development of teachers' epistemic identities. Repeated exposure to the epistemic core ideas with different examples for the aims and values, practices, methods and knowledge as applied to different chemistry examples (see Chap. 2) is likely to consolidate teachers' understanding about the epistemic core. Such exposure can potentially make the teachers more at ease with teaching the epistemic core, thereby addressing any affective or emotional constraints. CPD efforts can promote epistemic rationality by institutionalising rational criteria and standards in interactions among peers. Engagement in CPD activities design to promote the development of epistemic identities will necessarily position teachers in recognising their pivotal roles in enacting epistemic objectives.

## 8.4  Implications for Future Research

The content of the book raises questions for future research. Given the relatively recent history of the area of identity in science education research, "epistemic identity" is one that is ripe for further theoretical and empirical investigations. For example, from a theoretical perspective, the potential interplay of the cognitive, affective, social and institutional dimensions can be investigated. From an empirical perspective, research instruments and methodological tools can be designed to facilitate data collection and analysis on the various dimensions and their relationships. While the focus of our present work has been on pre-service teachers, a similar line of research can be conducted on the epistemic identities of in-service teachers and students. Overall, the construct of "epistemic identity" can be developed through both theoretical and empirical input to extend the research and development work presented in the book. This construct was something that we did not begin the book project with but rather we reached at towards the end of the process of engaging in the book (i.e. given our realisations as teacher educators through our engagement in the process of transformation of theoretical ideas into empirical and practical use). Hence our take on it is fairly limited and preliminary at this stage. Future research can be devoted to investigate the content, relevance and application of such a construct in chemistry teacher education more robustly.

When we embarked on this book project, we did not know what would be possible to accomplish with pre-service teachers. For example, considering the vast amount of literature on nature of science that reports the difficulties experienced by both in-service and pre-service teachers (e.g. Akerson, Buzelli, & Donnelly, 2010), we were mindful of the cognitive demands that we were placing on pre-service teachers with content that was unfamiliar to them. Hence, in the teacher education sessions presented in Chap. 4, we did not pursue a very nuanced approach to the epistemic core, for instance, through examples from history of chemistry. Rather, we provided some examples from typical school chemistry topics and engaged the pre-service teachers in the adaptation of the epistemic core to these examples. As a result, they produced their own adaptations of the epistemic core idea and represented them through the use of everyday and scientific analogies. However, examples of more advanced versions of the epistemic core idea were provided. For instance, in Chap. 2 examples of aims and values of chemistry such as "simplicity" were reflected in Hoffman, Minkin and Carpenter's (1997) discussion of the application of this value in the context of reaction mechanisms in chemistry.

Another further nuanced example was drawn about "objectivity" where "... objective methods are tied up with economic ends; they involve the combination of productivity and accuracy" (Baird, 2000, p. 100). The sense of intermixing of some traditional epistemic aims and values of science with other values such as productivity in the context of chemistry is further exemplified in the work of Bensaude-Vincent (2013) who discusses technoscience in its commercial and industrial context. Such extent and level of nuance would be something that could be followed up with particular case studies with further input as pre-service and in-service teachers become more versed in the epistemic aspects of chemistry. Without first having a sense of the meta-level categorisation of "aims and values" in the first instance, it is difficult to imagine how a more nuanced and discipline-specific articulation of any of the categories could have been pursued. Furthermore, the pre-service teachers will need to develop a sense of ownership of ideas as we observed in their visual representations in Chaps. 5 and 6. The use of everyday and scientific analogies had helped them make the ideas we were advancing in the sessions relevant. Future research can investigate optimal routes to teachers' learning of the epistemic core. For example, various contrasts of domain-general and domain-specific characterisations of the epistemic core could be taught and compared in terms of impact on pre-service teachers' understanding.

In discussing the epistemic core of chemistry, we capitalised on Erduran and Dagher's (2014) *Generative Images of Science* which were derived from a review of philosophy of science literature and proposed for pragmatic educational purposes. Erduran and Dagher defined nature of science as a cognitive-epistemic and social-institutional system. They produced a set of categories around the various aspects of this system and included a visual representation to illustrate how significant ideas around them can be captured visually. We focused only on the epistemic dimension because this is the focus of our book. The use of the images in the teacher education intervention raises numerous questions for future research that are based on the visual components of this line of research. For example, (a) How can visual representations of the epistemic core be assessed? What are the indicators of visual quality? (b) How do teachers' visual images of the epistemic core change as they develop epistemic identities? (c) How does the use of visual images of the epistemic core impact teachers' pedagogical content knowledge? Such questions can potentially inform teacher educators about how to make more use of visual imagery in their teaching practice.

Due to programme-specific constraints, it was not possible to observe the pre-service teachers during their teaching practice in the teacher education intervention described in Chaps. 4, 5 and 6. Thus, an important area for future research is the investigation of how pre-service teachers enact their representations and perceptions in teaching students in schools. For instance, what content on the epistemic core do pre-service teachers select to include in their lesson planning and why? How do pre-service chemistry teachers evaluate their students' understanding of the epistemic core? Apart from an emphasis on teaching practice, a further area of related work involves the role of the mentor teachers who are involved in supporting the pre-service teachers. If mentor teachers themselves are not familiar with the epistemic core, it is questionable how they may support the pre-service teachers. Hence, there's work to be done on the continuous professional development of mentors. Future research can focus on how these mentor teachers' own learning plays out in their mentorship roles.

The empirical component of the empirical research underpinning the book was conducted in a teacher education context. However the book raises questions about what can potentially be done in chemistry departments with undergraduate and potentially postgraduate students. Presumably understanding the epistemic dimensions of chemistry could be a broader goal of chemistry education than simply chemistry teacher education. As we noted in Chap. 3, there are some chemistry departments that offer courses on philosophy of chemistry to chemistry undergraduates (e.g. see http://www.hyle.org/service/courses.htm for a list of international courses). However these courses tend to be optional, and not integrated by design into chemistry teaching. If the audience of these chemistry courses also include future chemistry teachers, it would be beneficial for them to be introduced to the epistemic aspects of chemistry much earlier in their undergraduate chemistry courses so that the subject teaching itself becomes more epistemically robust. Earlier and repeated exposure to epistemic perspectives could also potentially foster undergraduate chemistry students' epistemic identities. It may also facilitate undergraduate chemistry students' understanding of difficult chemistry concepts because it would provide a chance to unpack and internalise these concepts. The sessions we outlined in Chap. 4 can easily be adopted for use in undergraduate chemistry courses. However, existing content in undergraduate chemistry education can also be reoriented to infuse some elements of the epistemic core without creating too many demands on the workload of the chemistry instructors who themselves will need to be versed in epistemic themes about chemistry. Empirical investigations can be carried out to test the impact of such infusion of the epistemic core in undergraduate chemistry courses.

## 8.5   Strengths and Limitations of the Book

The overall aims of this book were twofold: (a) to synthesise theoretical perspectives from philosophy of chemistry and teacher education, traditionally disparate bodies of literature, and (b) to explore how such perspectives can be infused into the design and implementation of teacher education programmes. Having synthesised the literatures on philosophy of chemistry and teacher education, we designed, implemented and evaluated a teacher education intervention. One of the strengths of the book is the demonstration of how the transformation of theoretical ideas into practically useful tools can be accomplished. For example, when we learn from the philosophical literature that there is a diversity of scientific methods and argue that this theme is an important one to capture in chemistry education (i.e. in light of traditional problems about the teaching of the scientific method), we make some progress. The appeal to the literature on philosophy of chemistry provides us with some clues about *what* to include in chemistry education, not *how* to include such content. Educators need some tools that can provide a focus for making theory transformation possible. The focus of the epistemic core based on Erduran and Dagher's (2014) work itself based on philosophy of science presented us with such focus. Furthermore the visual tools took us a step forward in capitalising on the power of visualisation in learning. We factored ourselves in the entire process as mediators of the theory-practice transformation recognising the importance of our own professional identities in enabling us to engage in this task. Our illustration of the processes that underpin the transformation of theory into practice, we believe, is going to provide a useful example for other educators who often are caught in the midst of theoretical perspectives and practical demands that are too wide to be bridged.

One of the key findings of the empirical component of the book concerns the nature of the pre-service teachers' images of the epistemic core. Despite decades of research on nature of science, the visual aspects are fairly limited in the literature (e.g. Erduran & Kaya, 2018). The study of visual images is not only about supporting learning but also it is about expressing and monitoring learning. The contrast of personal images versus disciplinary generative images (see Chap. 7) paves the way of future studies with other participants' (e.g. secondary students, in-service teachers) representations. *Personal Images of Science* may offer both practitioners and researchers alike new possibilities in addressing nature of science in teaching and learning. For example, they can be used as assessment tools to evaluate the participants' understanding and to consider the subsequent stages in teaching to move the participants towards disciplinary depictions. Researchers might be interested in investigating the progression in the quality of images as the participants engage more in learning environments that use such images. The criteria for evaluation might be related to the key themes intended for each image. For instance, in the case of scientific knowledge, whether or not an image consists of the idea of *growth* and *interdependence* could potentially serve as evaluation criteria.

While the aims of the book were addressed by undertaking the theoretical and empirical investigations reported, the results from pre-service teachers were limited to their individual outcomes in terms of their perceptions and representations. It was not possible to observe and analyse their teaching practice to investigate how they might have implemented the understanding that they developed in the context of chemistry lessons. Teachers' beliefs and knowledge may not necessarily be reflected in their teaching practice, and hence, the outcomes reported in Chaps. 5 and 6 should be taken within the context of their purpose intended in the book. In future studies, the teacher education intervention can be repeated at a different time in the academic year or in another national context to add a further dimension about the practical teaching aspects. A further limitation in the empirical dimension of the book is that while we provided much in-depth exploration of the pre-service teachers' enactment of the epistemic core of chemistry, the sample is small and the generalisability of the findings is limited. We should note, however, that the data presented in the book is consistent with the larger sample of the whole class of 14 pre-service teachers who took the module (see Erduran & Kaya, 2018 for further evidence.) Although the qualitative analyses were in line with the aims in the book, an even larger sample size of pre-service teachers would help to identify some trends in pre-service teachers' perceptions of the epistemic core. Our orientation in this book has been one of "proof of concept", i.e. to show that it is possible to infuse epistemic perspectives in chemistry teacher education and observe empirically the outcome through various methodological approaches such as visual and verbal accounts. Considering the vast amount of literature on how teachers' learn (see Chap. 3), we were aware that a teacher education model based on a transmission model of content of philosophy of chemistry would have had little benefit to the pre-service teachers. We designed and implemented a programme of work that assumed active learning strategies such as discussions and posters (see Chap. 4). Hence, our use of limited but in-depth qualitative data is fit for purpose in illustrating a tight, albeit limited, account of how to design, implement and assess a teacher education intervention on influencing pre-service teachers' epistemic thinking.

In bonding philosophical perspectives with chemistry teacher education, we have selected heuristics, strategies and ideas that, as science teacher educators and researchers, we anticipated utility and value for education. Our professional judgment as teacher educators working in specific national and cultural contexts suggested to us how we can design a realistic intervention that would be acceptable within the teacher education programmes we operate in. For example, due to the flexibility of elective coursework as well as the academic nature of the teacher education programme at Bogazici, we could include research perspectives along with practically useful pedagogical content in the sessions. A more decontextualised and philosophical content would not have been appropriate for the cohort of pre-service teachers in question. Hence, while our focus on the epistemic core was fairly narrow and did not include many other potential perspectives and studies from the philosophy of chemistry nor history of chemistry literature, it was accommodating to the level of students and institutional expectations. A philosopher of chemistry might find some of the perspectives reported in the book fairly simplistic. We believe that there is a balance to be struck between the philosophical and theoretical underpinnings of the issues that we are exploring and their pragmatic and realistic implementation in the actual contexts of teacher education for the purposes of practising teachers. In bridging the theoretical and practical worlds of philosophy of chemistry and teacher education, teacher educators have to be mindful of research evidence on how pre-service teachers learn and the constraints that they operate in.

Finally we should offer a note of caution in extrapolating the findings from the empirical investigations on pre-service teachers to students in school. In the teacher education intervention, the domain-general visual tools and themes were situated in chemistry examples, both in our own instructional input and also in pre-service teachers' own extensions. In this scenario, the pre-service teachers already had chemistry subject knowledge to draw from. In the case of secondary schooling, the same kind of strategy may not work. In other words, it would be unlikely that students can populate the epistemic core when they haven't yet learnt the chemistry concepts. Hence, the strategies reported in the book are restricted to the learning of pre-service teachers. It is not entirely unreasonable to expect that teachers' and students' learning will encompass different content and routes. Indeed, it is widely accepted that they do due to the different ways that these two cohorts of individuals relate to the subject. For example, chemistry teachers do not only need to learn about the subject knowledge from scratch given they have had some exposure at least at secondary level but rather they are challenged to transform this subject knowledge into teachable content (see Chap. 3). It is an entirely separate research agenda to investigate how the ideas and tools used in the teacher education intervention can be adopted for the teaching of students. The pre-service teachers in our project began to consider this mission by applying their learning to new chemistry topics (see Chap. 4). However further research is needed to investigate how students' learning of the epistemic core can be supported. An interesting area of research would be to examine the contrast of trajectories for students' versus teachers' learning of the epistemic core and how the domain-general versus domain-specific features play out in these trajectories.

## 8.6 Conclusions

In concluding the chapter and the book, it is worthwhile to contextualise the book in the broader landscape of chemistry education. Broadly speaking, the book is situated in what is typically referred to as "Chemistry Education Research" or CER. Herron and Nurrenbern (1999) defined CER as a "scholarship focused on understanding and improving chemistry learning". Bunce and Robinson (1997) had previously referred to CER as the "third branch of our profession" covering topics such as "how and why students learn", "why is chemistry difficult to learn" and "what facilitates effective chemistry teaching and learning". Some definitions of CER thus emphasise chemistry education as a pragmatic undertaking, an "add-on" to chemistry as a domain, and consider education in its practical utility. In contrast, this book is positioned from a social science perspective, drawing on interdisciplinary theoretical perspectives on philosophy of chemistry and educational research on teacher education, and methodological approaches in the social sciences such as interviews, analysis of visual imagery and self-study. In our approach, there is purposeful theorising and empirical testing of educational content and strategies, leading to academic knowledge about chemistry education. As such, the departure point for the rationale of the research is not chemistry as a domain but rather educational research.

It is furthermore worthy to situate the book in terms of its methodological and empirical approaches. Teo, Goh and Yeo (2014) reviewed the content of key journals for the coverage of research methods in chemistry education research over a 10-year period and observed that they used qualitative, quantitative and mixed methods. These researchers coded 650 papers in high-impact journals (e.g. CERP, IJSE) as qualitative, quantitative or mixed methods. The papers were coded as qualitative if only qualitative methods such as interviews and observations were used and the data were represented qualitatively (e.g. narratives and interview excerpts). The papers were coded as quantitative if only quantitative instruments such as surveys using Likert scale and tests with prescribed options were used. Additionally, the data were analysed using statistical methods and represented quantitatively showing numbers, charts, tables and so on. The papers were coded as mixed methods if a mixture of qualitative and quantitative instruments (e.g. questionnaires with Likert scale items and open-ended response) and methods were adopted. Additionally, the data were analysed and/or represented qualitatively and quantitatively. For example, qualitative interview transcripts may be transformed into quantitative data by counting the frequency with which a word or phrase was mentioned. Subsequently, the data were presented in tables and bar charts. Their findings show the number and percentage of papers that reported qualitative, quantitative or mixed methods studies. There were relatively more quantitative than qualitative studies. Most studies adopted mixed methods, and "mixing" typically occurred at the methods level with the use of qualitative and quantitative approaches such as interviews and Likert scale surveys. Additionally, many mixed methods studies involved the transformation of qualitative data into quantitative data by coding and counting the frequency. Overall, Teo et al. (2014) found that there is a general increasing trend for all three methods over the 10 years.

The empirical approach in the book was based on qualitative methodologies including interviews, analysis of visual imagery and self-study. Data were collected through structured interviews and used samples of pre-service teachers' drawings and verbal statements to make inferences about the impact of the intervention on their perceptions and representations. Our own reflections as teacher educators reported in Chap. 7 employed a self-study which constitutes yet another methodological approach. Given our overall aim of infusing the epistemic core of chemistry in teacher education, an in-depth exploration of the pre-service teachers' thinking and reasoning processes was important to capture. Overall, both the theoretical and empirical components of the research sit within a very limited research domain. In particular, based on Teo and colleagues' (2014) statistics, chemistry teacher education (i.e. 4.8%) and history and philosophy of chemistry in chemistry education empirical research (i.e. 0.6%) have been studied in fairly small proportions in chemistry education research. Hence, the intention is to contribute to these areas in a comprehensive fashion by addressing them in unison. Furthermore, the work is theoretical, empirical and pragmatic in nature.

The underlying assumption in the book concerns the recognition that philosophy of chemistry can potentially help improve chemistry education. As education specialists, we began by learning about philosophy of chemistry ourselves and exploring how we can apply what we learn in our teacher education endeavours. As teacher educators and researchers of teacher education, philosophy of chemistry is an area that is still fairly external to our daily routine. As we reviewed the literature, we were concerned about focusing on content that would be relevant and accessible to pre-service teachers. We were not interested in producing a course on philosophy of chemistry for pre-service teachers. Rather, we were driven by transforming some ideas from philosophy of chemistry to embed in the learning of pre-service teachers so as to encourage and support their epistemic thinking. The "epistemic core" idea and the associated visual tools from Erduran and Dagher's (2014) provided a fruitful mechanism to organise and structure our own thinking as teacher educators, and they constrained the boundaries of a vast literature so that we could design a module.

The teacher education intervention provided evidence on the outcomes of pre-service teachers' representations and perceptions of the epistemic core. Our reflections on the design, implementation and research processes made us recognise the significance of the construct of "identity" leading us to propose a framework on "epistemic identity" that can be investigated in future research and development efforts. The epistemic identity framework extends the focus on the cognitive elements of learning the epistemic core to be more comprehensive and inclusive of the other related dimensions of learning including the affective, social and institutional aspects. Overall, we have forged an interplay between the predominantly theoretical accounts underpinning the epistemic core and the empirical and practical accounts of teacher education. The book thus provides an example of how interdisciplinary theoretical frameworks (e.g. philosophy of chemistry and teacher education) can be transformed into pragmatic (e.g. content and delivery of teacher education) and research (e.g. analysis of verbal and visual data) outcomes.

In concluding the book, we reflect on a mindset in chemistry education represented in Kornhauser's words (1979) which still prevail to this day:

> The methods of chemical education are derived from the structure, logic and methods of chemistry itself. No other discipline can replace chemical science as the basis of the methodology of chemical education. (p. 32)

While chemistry as a domain is of course critical in chemistry education, so are meta-perspectives on chemistry as well as empirical evidence from educational research on how teachers learn to teach chemistry and how students learn chemistry. Chemistry education is a complex endeavour that appeals not only to disciplinary knowledge of chemistry but also to cognitive, epistemic and pedagogical processes, among others. Ignoring the role of the other foundational disciplines in informing chemistry education would essentially ignore evidence on what knowledge means in chemistry (i.e. meta-perspectives offered by philosophy of chemistry) and what it means to teach and learn chemistry (i.e. educational research including research on teacher education). With respect to the aims of this book in particular, coordinating theoretical and empirical perspectives is vital because (a) empirical evidence on how teachers learn can contribute to how understanding of chemistry can be supported, (b) designing and testing of teacher education strategies can help establish what is possible for teachers to learn and (c) theoretical accounts of how chemistry works help clarify how knowledge and knowledge development processes operate in chemistry. In conclusion, the book rests on an evidence-based approach, drawing on meta-perspectives on chemistry and transforming justified theoretical perspectives on the epistemic nature of chemistry into pragmatic approaches in teacher education. Empirical evidence is presented on the outcome of the implementation of such a transformation and some practical strategies are specified for the improvement of chemistry teacher education. Ultimately, the education of future generation of chemists as well as everyday citizens who are well informed about chemistry depends on teachers who understand what chemistry is about and who can apply such understanding in their teaching practice. Robust disciplinary knowledge coupled with reflective accounts on the nature of chemistry and chemistry teaching are likely to empower chemistry teachers for effective teaching.

# References

Akerson, V., Carter, I. S., & Elcan, N. (2016). On the nature of professional identity for nature of science. In L. Avraamidou (Ed.), *Studying science teacher identity* (pp. 89–110). Rotterdam, the Netherlands: Sense Publishers.

Akerson, V. L., Buzzelli, C. A., & Donnelly, L. A. (2010). On the nature of teaching nature of science: Preservice early childhood teachers' instruction in preschool and elementary settings. *Journal of Research in Science Teaching, 47,* 213–233.

Avraamidou, L. (2014). Studying science teacher identity: Current insights and future research directions. *Studies in Science Education, 50*(2), 145–179.

Baird, D. (2000). Analytical instrumentation and instrumental objectivity. In N. Bhushan & S. Rosenfeld (Eds.), *Of minds and molecules: New philosophical perspectives on chemistry* (pp. 90–113). Oxford, UK: Oxford University Press.

Beijaard, D., Meijer, P., & Verloop, N. (2004). Reconsidering research on teachers' professional identity. *Teaching and Teacher Education, 20,* 107–128.

Bensaude-Vincent, B. (2013). Chemistry as a technoscience? In J. P. Llored (Ed.), *The philosophy of chemistry: Practices, methodologies and concepts* (pp. 330–341). Newcastle upon Tyne, UK: Cambridge Scholars Publishing.

Bianchini, J. A., Cavazos, L. M., & Helms, J. V. (2000). From professional lives to inclusive practice: Science teachers and scientists' views of gender and ethnicity in science education. *Journal of Research in Science Teaching, 37*(6), 511–547.

Broggy, J., O'Reilly, J., & Erduran, S. (2017). Interdisciplinarity and science education. In B. Akpan & K. Taber (Eds.), *Science education: An international course companion* (pp. 81–90). Rotterdam, the Netherlands: Sense Publishers.

Brun, G., & Kuenzle, D. (2008). A new role for emotions in epistemology? In G. Brun, U. Doguoglu, & D. Kuenzle (Eds.), *Epistemology and emotions* (pp. 1–32). Aldershot, UK: Ashgate.

Bunce, D. M., & Robinson, W. R. (1997). Research in chemical education – the third branch of our profession. *Journal of Chemical Education, 74*(9), 1076–1079.

Carlone, H. B., & Johnson, A. (2007). Understanding the science experiences of women of color: Science identity as an analytic lens. *Journal of Research in Science Teaching, 44*(8), 1187–1218.

Demerath, L. (2006). Epistemological identity theory: Reconceptualizing commitment as self-knowledge. *Sociological Spectrum, 26,* 491–517.

Ellis, V., Blake, A., McNicholl, J., & McNally, J. (2011). *The work of teacher education: The final research report for the Higher Education Academy.* Subject Centre for Education, ESCalate.

Erduran, S., & Dagher, Z. R. (2014). *Reconceptualizing the nature of science for science education: Scientific knowledge, practices and other family categories.* Dordrecht, the Netherlands: Springer.

Erduran, S., & Kaya, E. (2018). Drawing nature of science in pre-service science teacher education: Epistemic insight through visual representations. *Research in Science Education.* https://doi.org/10.1007/s11165-018-9773-0

Foster, D. (2018). *Initial teacher training in England.* London: House of Commons, UK parliament.

Green, J. A., Sandoval, W. A., & Braten, I. (2016). *Handbook of epistemic cognition.* London/New York: Routledge.

Helms, J. V. (1998). Science – and me: Subject matter and identity in secondary school science teachers. *Journal of Research in Science Teaching, 35*(7), 811–834.

Herron, J. D., & Nurrenbern, S. C. (1999). Chemical education research: Improving chemistry learning. *Journal of Chemical Education, 76*(10), 1354–1361.

Hoffman, R., Minkin, V. I., & Carpenter, B. K. (1997). Ockham's razor and chemistry. *Hyle – An International Journal for the Philosophy of Chemistry, 3,* 3–28.

Hogg, M. A., & Abrams, D. (1988). *Social identifications: A social psychology of intergroup relations and group processes.* London: Routledge.

Kelly, T. (2003). Epistemic rationality as instrumental rationality: A critique. *Philosophy and Phenomenological Research, LXVI*(3), 612–640.

Kornhauser, A. (1979). Trends in research in chemical education. *European Journal of Science Education, 1*(1), 21–50.

Luehmann, A. L. (2007). Identity development as a lens to science teacher preparation. *Science Education, 91*(5), 822–839.

Maher, F. A., & Tetreault, M. K. T. (2001). *The feminist classroom: Dynamics of gender, race, and privilege.* Lanham, MD: Rowman & Littlefield Publishers, Inc..

Murray, J. (2008). Towards a re-articulation of the work of teacher educators in higher education institutions in England. *European Journal of Teacher Education, 31*(1), 17–34.

Murray, J., & Kosnik, C. (2011). Academic work and identities in teacher education. *Journal of Education for Teaching: International Research and Pedagogy, 37*(3), 243–246.

Murray, J., & Male, T. (2005). Becoming a teacher educator: Evidence from the field. *Teaching and Teacher Education, 21*(2), 125–142.

Peters, E. E., & Kitsantas, A. (2010). Self-regulation of student epistemic thinking in science: The role of metacognitive prompts. *Educational Psychology, 30*(1), 27–52. https://doi.org/10.1080/01443410903353294

Rodgers, C. R., & Scott, K. S. (2008). The development of the personal self and professional identity in learning to each. In M. Cochran-Smith, S. Feiman-Nemser, & D. J. McIntyre (Eds.), *Handbook of research on teacher education: Enduring questions in changing contexts* (pp. 732–755). New York: Routledge.

Roth, W.-M., & Tobin, K. (Eds.). (2007). *Science, learning, identity: Sociocultural and cultural-historical perspectives.* Rotterdam, the Netherlands: Sense Publishers.

Sutherland, L., Howard, S., & Markauskaite, L. (2010). Professional identity creation: Examining the development of beginning preservice teachers' understandings of their work as teachers. *Teaching and Teacher Education, 26*, 455–465.

Swennen, A., Jones, K., & Volman, M. (2010). Teacher educators: Their identities, sub-identities and implications for professional development. *Professional Development in Education, 36*(1–2), 131–148.

Teo, T. W., Goh, M. T., & Yeo, L. W. (2014). Chemistry education research trends: 2004–2013. *Chemistry Education Research and Practice, 15*, 470–487.

Thoits, P. A. (1986). Multiple identities: Examining gender and marital status differences in distress. *American Sociological Review, 51*(2), 259–272.

Turner, J. C., Hogg, M. A., Oakes, P. J., Reicher, S. D., & Wetherell, M. S. (1987). *Rediscovering the social group: A self-categorization theory.* New York: Basil Blackwell.

Volkmann, M. J., & Anderson, M. A. (1998). Creating professional identity: Dilemmas and metaphors of a first-year chemistry teacher. *Science Education, 82*(3), 293–310.

Wenger, E. (1998). *Communities of practice: Learning, meaning, and identity.* New York: Cambridge University Press.

Zembal-Saul, C. (2016). Implications of framing teacher development as identity construction for science teacher education research and practice. In L. Avraamidou (Ed.), *Studying science teacher identity* (pp. 321–332). Rotterdam, the Netherlands: Sense Publishers.

Printed in the United States
By Bookmasters